HABITATS AND THE ENVIRONMENT

INVESTIGATIONS

Peter Freeland

Hodder & Stoughton

A MEMBER OF THE HODDER HEADLINE GROUP

Preface

The introduction of options into A Level Biology syllabuses increases the scope for practical investigations. This book contains 27 suggestions for investigations into various aspects of ecology. These proposals have been tested on a group of sixth-formers, who have used them either as class practicals or as a starting point for individual problem-solving investigations. Wherever possible, the investigations make use of simple, inexpensive apparatus and chemical compounds currently believed to present no threat to the health or safety of pupils.

The general lay-out of each investigation is as follows.

Background information – a statement of the aims, purpose or relevance of the investigation
Preparation – instructions for teachers and laboratory assistants in laboratory-based exercises
Equipment and materials – a list of all the requirements
Experimental procedure – practical instructions aimed at pupils
Extension work – proposals for further investigations on the same or a closely related topic.

As some investigations use unfamiliar materials, a list of suppliers is included as an appendix. This list however is by no means comprehensive and does not necessarily contain the only, or most economical, supplier of materials.

I hope pupils and teachers will find this book a useful resource for practical work.

PWF 1991

The following titles are also available in the *FOCUS ON BIOLOGY* series:

Habitats and the Environment
ISBN 0 340 53267 x

Micro-organisms in Action
ISBN 0 340 53268 8

Micro-organisms in Action Investigations (pack)
ISBN 0 340 53922 4

Genetics and Evolution
ISBN 0 340 53266 1

Plant Science in Action
ISBN 0 340 60160 0

Plant Science in Action Investigations
ISBN 0 340 60099 3

Acknowledgements
I am most grateful to my laboratory assistant, Mrs M Moore, for preparing many of the investigations contained in this book, and to my students, both past and present, for their part in obtaining results.

PWF

Contents

Safety recommendations

To conform with the Health and Safety at Work Act (1974) it is important to adhere to the general and specific guidelines laid down in the following publications:

Micro-biology: An HMI Guide for Schools and Non-advanced Further Education HMSO
Hazardous Chemicals: A Manual for Schools and Colleges SSSERC
Hazards in the Chemical Laboratory (4th edition) The Chemical Society
Topics in Safety ASE

At appropriate points throughout these investigations, attention is drawn to potential risks by the use of hazard symbols. These are as follows.

Toxic compound

Corrosive compound

Inflammable compound

Wear eye protection
Goggles must be worn whenever a Bunsen burner is used

Wear plastic gloves
The major risks are of allergic reactions to enzyme powders and soil samples, or the absorption of toxic compounds through wet skin

Finally, as some LEAs have their own recommendations concerning the practical work used in their schools, teachers should always check with their advisers before attempting any unfamiliar procedure.

investigation 1

Recording data

Every large ecosystem has a living or biotic component, made up of many different species which function as producers, consumers and decomposers. Describing all the species in a wood or pond is a very difficult task, because some are not present throughout the year, while others are difficult to catch or identify. By using a combination of keys and picture books to identify specimens, it is usually possible to produce a fairly comprehensive list of species. This not only highlights the major groups of organisms that are present in an ecosystem, but also draws attention to those that are sparse or absent. Furthermore, records of this type often draw attention to variations between the flora and fauna of different woods, or ponds, in the same locality. Such differences can serve as a basis for further investigations.

Equipment and materials

- data sheets
- keys, picture books (for identification of species)
- clipboard
- pencil
- sweep nets
- pitfall traps
- small mammal traps

Experimental procedure

1 Prepare data sheets, as indicated in Table 1 overleaf. You may find it necessary to spread the headings over two or more pages, depending on the number of species that are present in an ecosystem. (A comprehensive, rational system of classification can be found in *Biological Nomenclature*, a publication sponsored by The Association for Science Education and the Institute of Biology.)

GROUP (Kingdom, Phylum or Class)	SPECIES	COMMENTS
PROTOCTISTA		
CHLOROPHYTA (green algae)		
PHAEOPHYTA (brown algae)		
RHODOPHYTA (red algae)		
BRYOPHYTA		
HEPATICAE (liverworts)		
MUSCI (mosses)		
FILICINOPHYTA (ferns)		
ANGIOSPERMATOPHYTA (flowering plants)		
Herbs		
Shrubs		
Sub-canopy trees		
Canopy trees		
FUNGI		
PLATYHELMINTHES (flatworms)		
ANNELIDA (earthworms)		
MOLLUSCA (snails)		
ARTHROPODA		
Crustacea (woodlice)		
Insecta		
Chilopoda (centipedes)		
Arachnida (spiders)		
CHORDATA		
Osteichthyes (bony fish)		
Amphibia		
Reptilia		
Aves (birds)		
Mammalia		

Table 1 List of species present in a habitat or ecosystem

2 Identify as many species as possible by using keys and picture books. List the species you have identified under their appropriate group. Wherever possible, add comments, such as those given in Table 2.

GROUP	SPECIES	COMMENTS
ANGIOSPERMATOPHYTA		
Shrubs	Ivy	Epiphytic on trees
ARTHROPODA		
Crustacea	Woodlouse	Detritivore, mainly in leaf litter
CHORDATA		
Mammalia	Common shrew	Small carnivore, eats insects, woodlice and snails

Table 2 Sample entries (from woodland)

investigation 2

Analysing plant cover

The description and analysis of plant cover at ground level often involves measurements of density, cover and frequency. All of these measurements may be made by using a wooden or metal frame called a **quadrat** (Figure 1). Conventionally, a quadrat is a 1m × 1 m square, subdivided into 10 cm × 10 cm squares, but both the size and shape can be varied for particular tasks. **Density** is a measure of the number of plants present within a quadrat. **Cover value** is an estimate of the approximate area covered by a given species, usually expressed as a percentage of the total area. **Frequency** is a measure of the probability of finding a plant within any given area. Estimates of density, cover value and frequency can be made in a lawn containing daisy plants or other weed species.

Equipment and materials

- daisy plants in a lawn
- 1 m^2 quadrat

Experimental procedure

1 Throw the quadrat to fall over a part of the lawn containing daisy plants. Count the number of daisy plants in the frame. This figure gives the density of daisy plants within a single quadrat.
2 Repeat procedure **1** either 10 or 25 times. Calculate mean values to give a more accurate estimate of plant density.
3 Again repeat prodecure **1**, but this time count the number of 10 cm^2 squares in which one or more leaves of daisy plants occur. This gives an estimate of shoot cover value. Repeat the procedure either 10 or 25 times, then calculate mean values to give a more accurate estimate of cover value.
4 Throw the quadrat at random and find out if there are any daisy plants within the area bounded by the frame after it has landed. Make 25 or 50 similar throws. Suppose after 50 throws, one or more daisy plants were found within 27/50 quadrats:

$$\text{frequency} = \frac{27}{50} \times 100 = 54\%$$

5 Calculate the frequency of daisy plants in the lawn.
6 Plot a histogram to compare estimates of density, shoot cover value and frequency. Evaluate the usefulness of each type of measurement.

Taking it further

1 Find out how estimates of plant density vary with the number of samples taken. Calculate the density of plant cover after 1, 5, 10, 20, 30 and 50 samples have been taken. Plot your results as a graph.
2 Use 1 m^2, 0.5 m^2 and 0.25 m^2 quadrats to find out how quadrat size affects estimates of plant density, cover value and frequency.

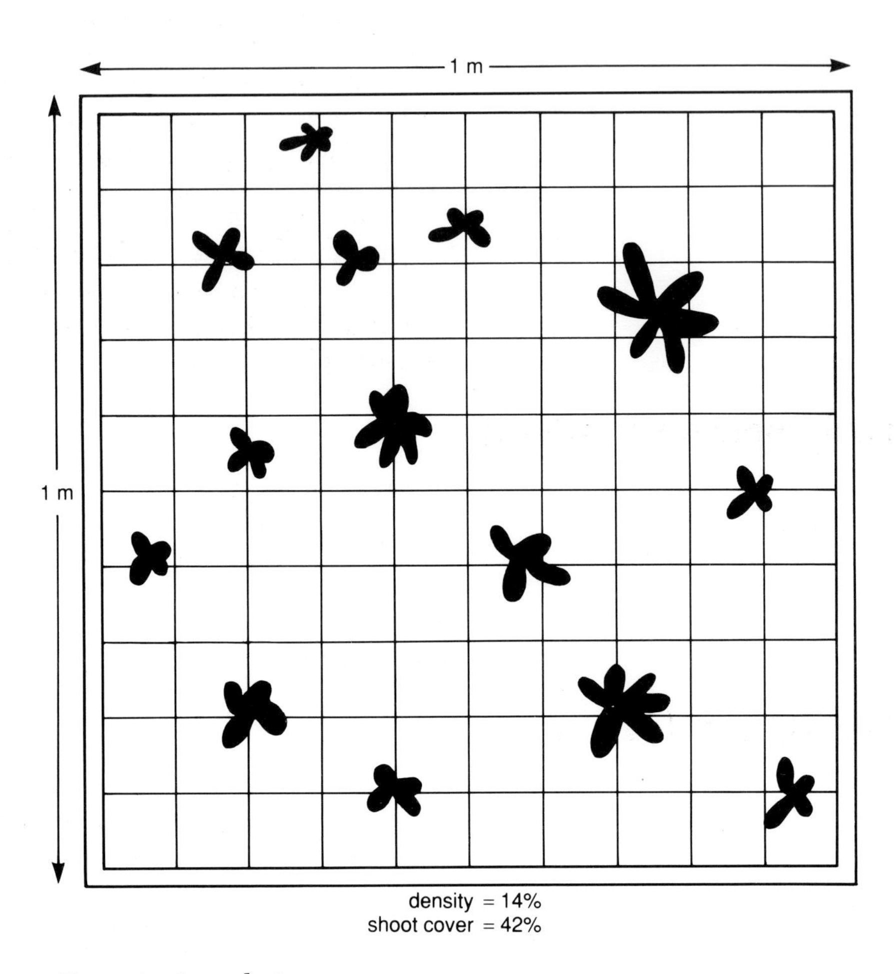

Figure 1 A quadrat

investigation 3

Analysing algae and lichen distribution

Plants that grow on tree trunks are called **epiphytes**. Some of the most common and widely distributed species include the powdery, light green alga called *Pleurococcus*, and many different species of leaf-like lichens. The distribution of these plants is determined, in part, by environmental factors, notably the direction of the wind and light intensity.

A suitable type of quadrat for sampling small epiphytes on tree trunks consists of a sheet of transparent plastic, 12 cm^2, available from suppliers (Figure 1). This quadrat is pinned at a fixed height above ground level on north-facing, east-facing, south-facing and west-facing surfaces of the trunk. The density of plants at each site is determined. Histograms are plotted to show the relationship between plant density and aspect.

Equipment and materials

- tree trunk bearing *Pleurococcus* and lichens
- 12 cm^2 plastic quadrat
- metre rule
- string
- compass
- drawing pins

Experimental procedure

1 Select a part of the trunk that is covered by *Pleurococcus* and lichens. Use the metre rule to measure the height above ground level. Tie string around the trunk at this level to mark off uniform height.
2 Use the compass to locate the magnetic north. Pin the quadrat in a north-facing position, with its upper margin in contact with the string.
3 Count the number of squares in which (a) *Pleurococcus* and (b) lichens occur. Express the density of each epiphyte as a percentage of the total number of squares in which counts have been made.
4 Repeat procedure **3** when the quadrat has been pinned in (a) east-facing, (b) south-facing and (c) west-facing positions.
5 Draw histograms to show how the distribution of (a) *Pleurococcus* and (b) lichens varies according to aspect. Explain your results.

Taking it further

1 Survey the distribution of *Pleurococcus* on a tree trunk from ground level to a height of 2 m. Plot your results as a histogram. Explain your results.

2 Use circles on the quadrat to determine the annual growth rate of leaf-like lichens or mosses. Alternatively, use the quadrat to determine the growth rate of shoots of ivy, an epiphytic flowering plant.

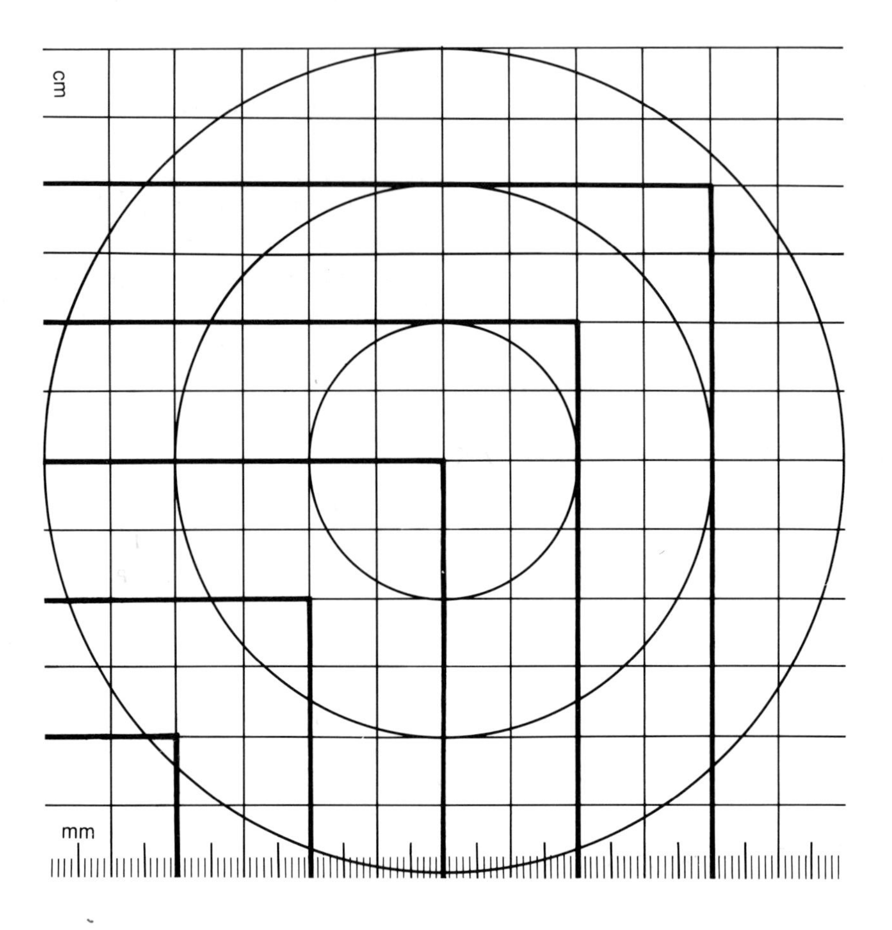

***Figure 1** Translucent plastic quadrat for sampling the surface of tree trunks, walls etc.*

investigation 4

Line and belt transects

A **transect** is an accurate plot of the distribution and arrangement of species along a straight line, or parallel-sided strip, of ground surface. **Line transects** are plotted from a straight line of string, supported by ranging posts. At regular intervals of 1.0 m, 0.5 m, 0.25 m or 10 cm along the string, organisms immediately beneath each mark are identified and their height recorded. Line transects are particularly useful where plants are growing on a gradient and there is a marked change in the vegetation from one zone to another (Figure 1). Such transitions may occur, for example, across a hedge, path or roadside verge. **Belt transects** are used to describe the vegetation, or distribution of animals, across a strip of vegetation bounded by parallel lines. The sea shore offers a particularly suitable situation for belt transects, which can be used to describe the distribution of algae or shelled molluscs such as periwinkles.

Equipment and materials

- low-growing hedge, flanked by a ditch (or zoned vegetation)
- transect line, marked in units of 10 cm
- 2 ranging posts (with alternate 50 cm red and white sections)
- string
- mallet
- spirit level
- metre rule
- flora
- graph paper

Experimental procedure

1 Drive the metal tips of the ranging posts into the ground with the mallet, one post on each side of the hedge.
2 Tie the transect line between the posts, and level it using the spirit level. Measure the distances H_1, H_2 and d. Calculate the percentage gradient (gradient %) from the following formula:

$$\frac{H_1 - H_2}{d} \times 100$$

(If the top of the hedge is more than 1 m above ground level, tie a piece of string at a distance of 1 m below the transect line, and measure heights from this lower line.)
3 Place the metre rule, in turn, at each 10 cm mark across the transect line. Record (a) the distance of the transect line from the ground, (b) the distance from the transect line to the top of vegetation, and (c) the species of plant immediately beneath each 10 cm mark.

4 On returning to the laboratory, make an accurate scale drawing on graph paper of a transect across the vegetation you have studied. Represent each species of plant by a realistic sketch, drawn to scale. Label the species and indicate the percentage gradient.

Taking it further

1 Set up parallel transect lines, at a height of 10–20 cm above ground level, and 1 m apart. Use a quadrat to determine the density of plants across the path. Plot a transect on graph paper to show how the density of species, and the height of individual plants, changes in the well trodden, lightly trodden and untrodden parts of the path.
2 Set up a belt transect across a sandy or rocky sea shore. Identify the species of (a) algae or (b) shelled molluscs across the transect. Use a quadrat to estimate their density. Make a scale drawing of the transect on graph paper.

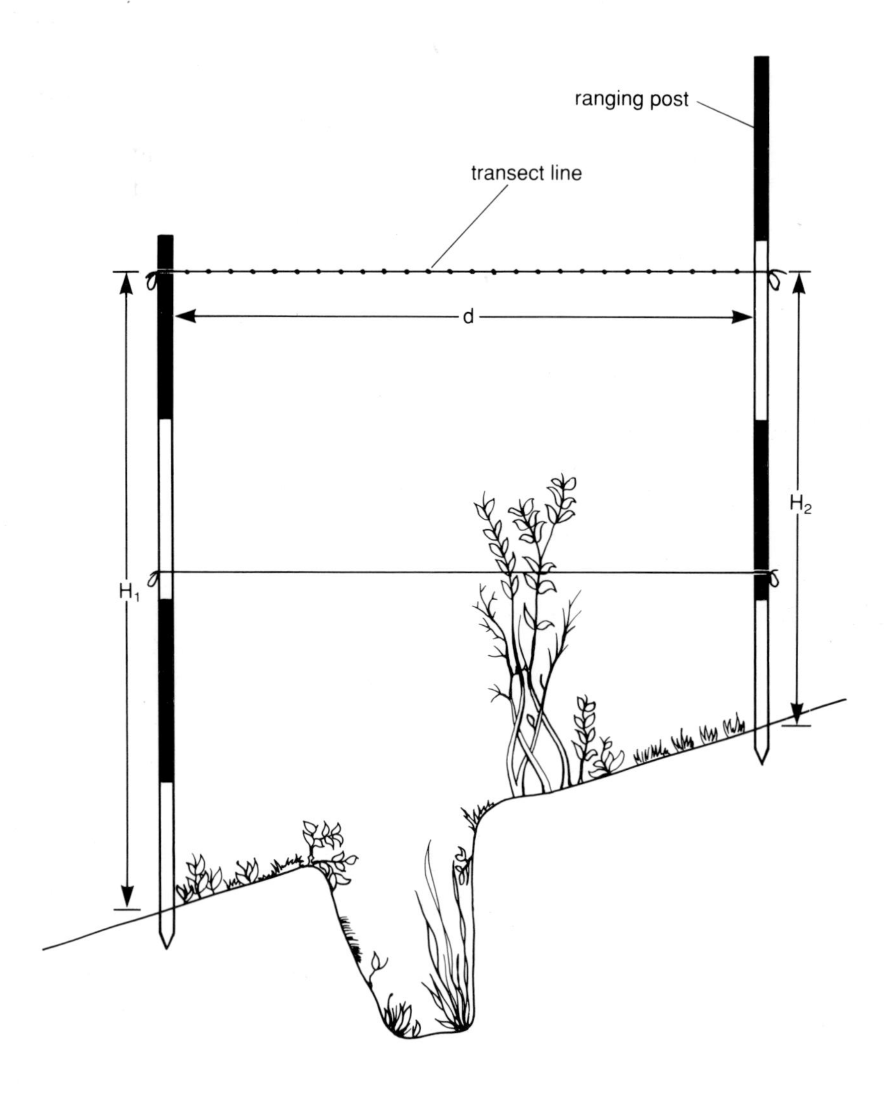

Figure 1 Transect across a low-growing hedge, with flanking ditch

investigation 5

Water, air and humus in the soil

Soil is a mixture of inorganic material (clay, silt, sand, gravel particles), non-living organic matter (humus), and living organisms (biomass). The spaces between these solid components are occupied by air or by soil solution. Amounts of water, air and humus in soils are extremely variable, depending partly on the nature of the overlying vegetation, partly on the extent to which the soil has been turned by animals or humans, and partly on the depth from which the soil sample is taken. In this investigation, simple techniques are used to estimate percentages, by volume or mass, of water, air and organic matter (humus) in soil samples.

Equipment and materials

- woodland soil
- garden soil
- 3 × 100 cm^3 measuring cylinders
- 2 × metal dishes
- 2 × rubber bungs (to fit measuring cylinders)
- hand trowel
- plastic bags and string
- Bunsen burner, tripod and sand tray
- spatula
- top pan balance
- incubator or oven, maintained at 60°C
- eye protection
- glass-marking pen

Experimental procedure

1 Select suitable woodland and garden sites.
2 Dig up 200–500 g of soil from each site. Put each sample into a plastic bag, tie it with string and label it.

Determining water content

3 On returning to the laboratory, weigh out 200 g of each soil. Weigh each metal dish and record its mass. Put each soil sample into a metal dish. Label the dishes '1' (woodland soil) and '2' (garden soil).
4 Transfer the dishes to an incubator or oven at 60°C. Heat the soil samples until their mass is constant.
5 Re-weigh each soil sample and record its mass. Calculate the percentage water in each sample. Substitute your value for the dry mass of soil (x) in the equation below.

Original mass of soil	200 g
Dry mass of soil	x g
Loss in mass	(200 – x g)
% water in sample	$(200-x) \times \frac{100}{200}$

Determining air content

6 Take two 100 cm^3 measuring cylinders and number them '1' and '2'. Fill cylinder 1 to the 50 cm^3 mark with dried woodland soil, and cylinder 2 to the 50 cm^3 mark with dried garden soil. Add 50 cm^3 tap water to each cylinder (Figure 1a). Fit a rubber bung to each cylinder, shake the contents, and allow the mixtures to settle.
7 Read and record the water level in each cylinder (level y). Calculate the percentage of air in each sample from the following equation:

$$(100 - y) \times \frac{1}{50} \times 100$$

Determining humus and mineral particle content

8 Add water to each measuring cylinder to the 100 cm^3 mark. Fit the bungs and shake the cylinders. After the soil and water have been thoroughly mixed, stand each cylinder on the bench and allow the particles to settle. Within a few minutes the soil particles in the measuring cylinder should have separated, as shown in Figure 1b. Estimate the percentage of humus, clay, sand and gravel in each soil sample. Record your results.

Determining humus content

9 Put on your eye protection. Weigh out 20 g of dried woodland soil on the sand tray. Stand the sand tray over a tripod and heat it with the Bunsen burner for 10–15 minutes (Figure 1c). Wait until the soil has cooled, then re-weigh.
10 Calculate the percentage of organic matter present.
11 Repeat procedures **9** and **10** using dried garden soil.
12 Tabulate your results for the organic content of both soils. Include estimates for humus (volume) and organic matter (mass). Which estimate is the most useful?

Taking it further

1 Devise and use a method for finding out if garden soil contains more organic matter in November than in June. Attempt to account for any differences you record. How might your results be affected by horticultural practices designed to maintain soil fertility?
2 Investigate how the water, air and humus content of woodland soil varies with depth, from the surface to, for example, a depth of 2 m. Present all your results in the form of histograms. Extend your investigation to include measurements of water, air and humus content at different seasons of the year.

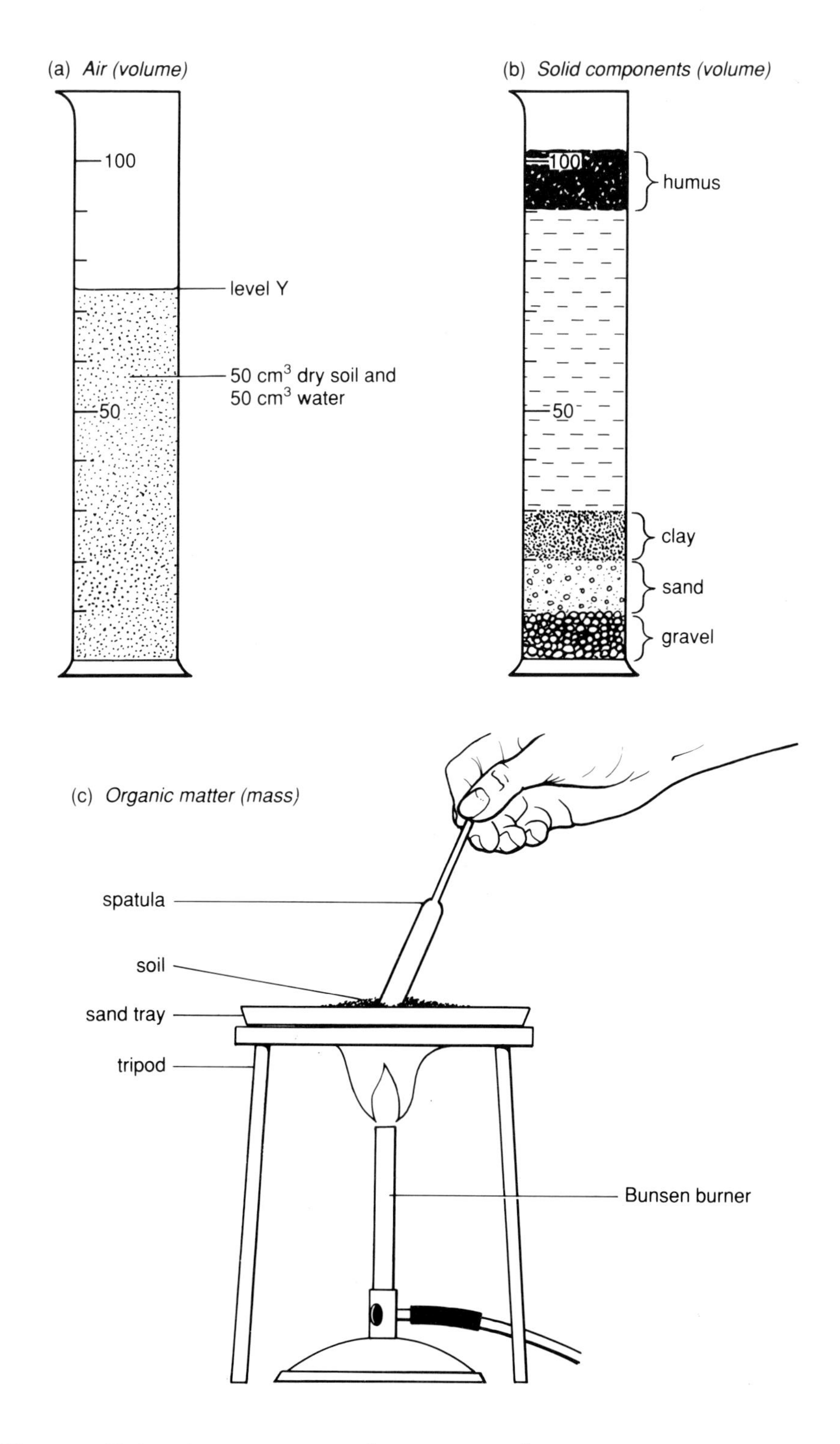

Figure 1 ***Estimating percentage volume or mass of components in soil samples***

investigation 6

Effects of lime on different soils

Lime, a general term for alkaline calcium salts (notably calcium oxide, hydroxide, carbonate or 'limestone'), is added to cultivated soils and pastures to neutralise acidity and to supply calcium (Ca^{2+}) ions. It also generally promotes the development of a soil microflora which oxidises organic matter and releases plant nutrients. An important chemical effect of adding lime to a soil is to release sodium (Na^+) and potassium (K^+) ions, bound to colloids, into the soil solution.

The addition of lime to clay causes the particles to aggregate forming larger particles, thus making the soil more permeable. It also promotes the formation of calcium-containing colloids, which are much less sticky and water-retaining than sodium-containing colloids. Lime, added to sand, acts as a weak cementing agent, increasing the cohesion of particles. This reduces the permeability of sand and increases its water-retaining capacity. The object of this investigation is to find out how the addition of lime to (a) clay, (b) sand and (c) loam affects the chemical and physical properties of these soils.

GENERAL TECHNIQUES

Equipment and materials

- distilled water
- 155 g dry clay
- 155 g dry sand
- 5 g dry loam
- 10 g lime
- universal indicator
- universal indicator colour chart
- Merckoquant Na^+-sensitive reagent sticks
- Merckoquant K^+-sensitive reagent sticks
- 7 × 100 cm^3 measuring cylinders
- 8 × test tubes in a rack
- 6 × filter funnels
- 6 × filter papers
- top pan balance
- clock
- ruler
- spatula
- glass-marking pen

Experimental procedure

Effect on pH

1 Weigh out three 50 g samples of clay. Add 0.5 g of lime to one sample and 2.5 g to another. Use the spatula to mix thoroughly the clay and lime.
2 Repeat procedure **1** with sand.
3 Draw three horizontal lines, approximately 1 cm apart, from the base of each of six test tubes. Number the tubes from **1–6**. Add the

following soils or soil/lime mixtures to the level of the lowest mark on each tube.

Tube number	*Addition*
1	clay
2	sand
3	clay + 0.5 g lime
4	sand + 0.5 g lime
5	clay + 2.5 g lime
6	sand + 2.5 g lime

Add distilled water to the second mark and universal indicator to the third (Figure 1a). Gently shake the mixtures, and allow particles to settle before reading pH values from the universal indicator chart. Record the pH of each soil solution. What is the effect of adding lime? Approximately how much lime should be added to (a) 50 g clay and (b) 50 g sand to bring their pH to within the range 5.5–7.0?

Effect on porosity and water retention

4 Take six measuring cylinders and number them from 1–6. Set up the filter funnels, lined with filter paper, above the measuring cylinders.
5 Weigh out 50 g samples of sand, clay and each of the mixtures containing 0.5 g lime and 2.5 g lime. Transfer each soil sample to a filter funnel above an appropriately numbered measuring cylinder (Figure 1b).
6 Pour 150 cm^3 tap water over each soil sample and record each of the following:
(a) the volume of water that collects in the measuring cylinder,
(b) the time taken for water to drain from the soil surface.
7 Calculate the percentage water retention by each soil sample. What do you conclude about the effects of lime on water retention by (a) clay and (b) sand?

Effect on soil solution

8 Put 2 g loam soil into each of the remaining test tubes. Number these tubes 7 and 8. Add 10 cm^3 distilled water to each tube and 0.5 g lime to tube 8. Shake the contents of each tube and allow the particles to settle. Test each soil solution with a Merckoquant Na^+- and K^+-sensitive reagent stick (Figure 1c). Record your results. Does the addition of lime have any effect on the concentration of Na^+ and K^+ ions in the soil solution?

Taking it further

1 Apply lime to a unit area of garden soil, about 5 m^2, in January–March, and dig it in; leave a similar area unlimed. In June–August compare the arthropod and annelid populations in treated and untreated areas of the garden. Extend the investigation to include measurements of pH, humus, water and air content in the treated and untreated areas.
2 Investigate the effect of (a) lime and (b) calcium nitrate (nitrochalk) on the growth of a lawn. Subdivide the lawn into three equal portions. Leave one portion untreated. Apply a surface

layer of lime to the second portion and nitrochalk to the third. At fortnightly intervals from April–September, mow each portion of the lawn and retain the mowings. Weigh the mowings from each part of the lawn and record their wet mass. At the end of September, present all your results in an appropriate graphic form. Summarise and evaluate the effects of each treatment.

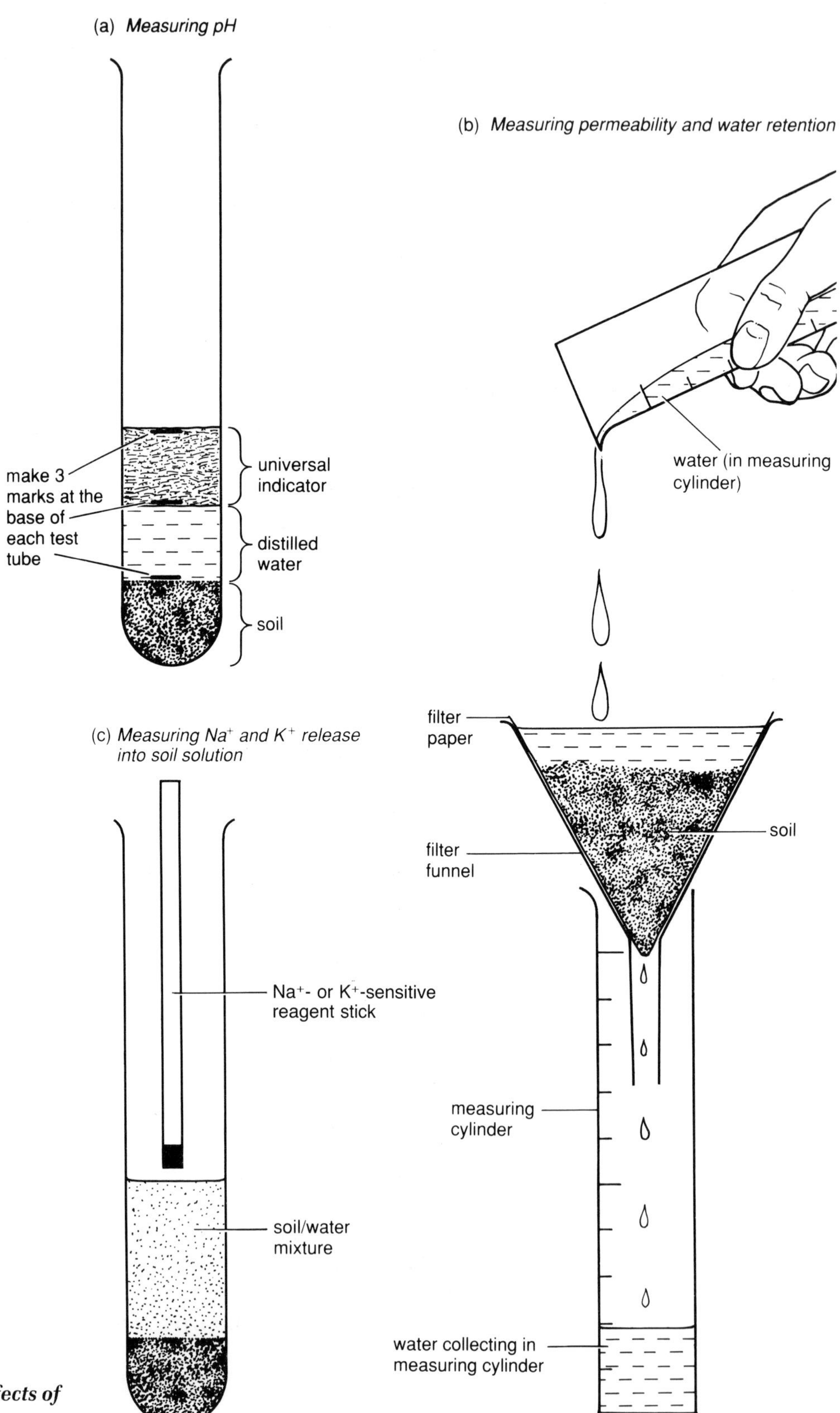

Figure 1 Investigating the effects of liming soil

investigation 7

Variation within species

Variation among individuals of the same species is the factor on which the forces of natural selection act in determining the course of evolution. The study of specimens in the field, including measurements of mass and length, can illustrate the extent to which the individuals of some species differ from one another. For instance, the leaves and flowers of many plant species often show marked variations. Similar variations in size, shape and colour can be found amongst molluscs on the sea shore. In this investigation you will examine shelled molluscs on a sea shore, measure them, weigh them and then plot histograms to show the number of individuals that fall within unit categories of mass, length and colour.

Equipment and materials

- shelled molluscs (e.g. dog whelk, periwinkle or common limpet)
- 2 × transect lines (50 m) with pegs
- 100 g spring balance, graduated in 1 g units
- plastic envelope or bag
- clip board
- paper and pencil
- ruler

Experimental procedure

1 Lay out two transect lines across a rocky or sandy shore, parallel to one another and 1 m apart.
2 Attach the plastic envelope or bag to the hook of the spring balance so it can serve as a receptacle for molluscs during weighing (Figure 1).
3 Select the most abundant mollusc, such as dog whelks. Start at one end of the transect and work along it, handling each mollusc in turn. For each mollusc, measure and record the following:
(a) mass,
(b) shell length,
(c) shell width,
(d) aperture length,
(e) aperture width.
4 In addition, record the number of individuals in which the shell is (a) white, (b) mottled brown on white, (c) banded brown, and (d) banded black. (If a different mollusc has been selected, devise your own categories of colouring.) Carefully return each mollusc to the position from which it was taken.
5 On returning to the laboratory, plot histograms to show the number of individuals that fall within given units of mass, length and categories of colouring. Figure 2 illustrates specimen results.

Taking it further

1 Collect 250 or more intact flowers of buttercup. Count the number of petals and sepals in each flower. Plot histograms to show variation in the numbers of petals and sepals.

2 Remove the leaves from a one-year-old twig of privet. Measure and record the length of each leaf. Plot a histogram to show variation in leaf length. Explain your results.

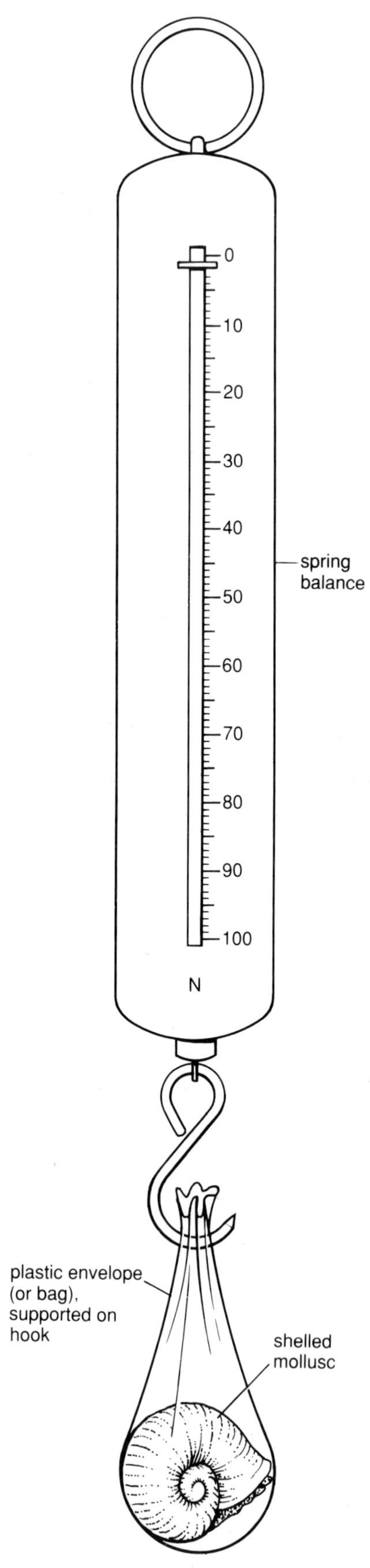

Figure 1 Weighing a mollusc with a spring balance

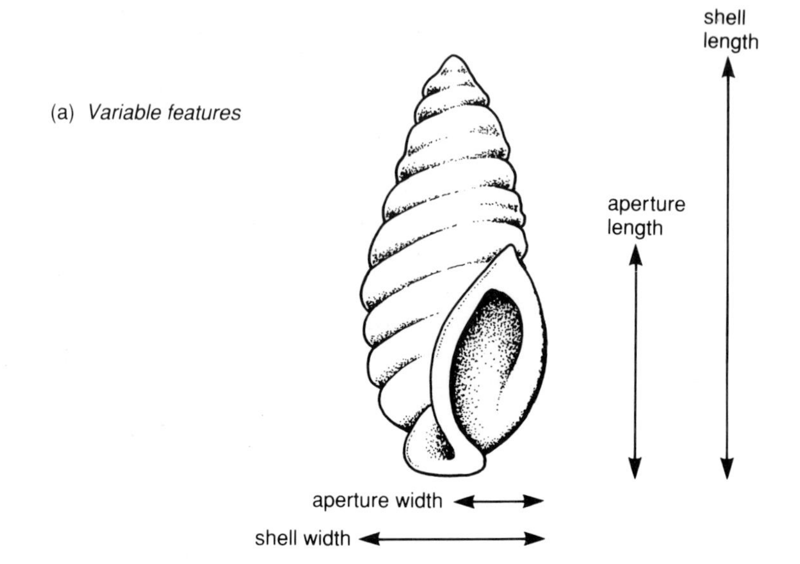

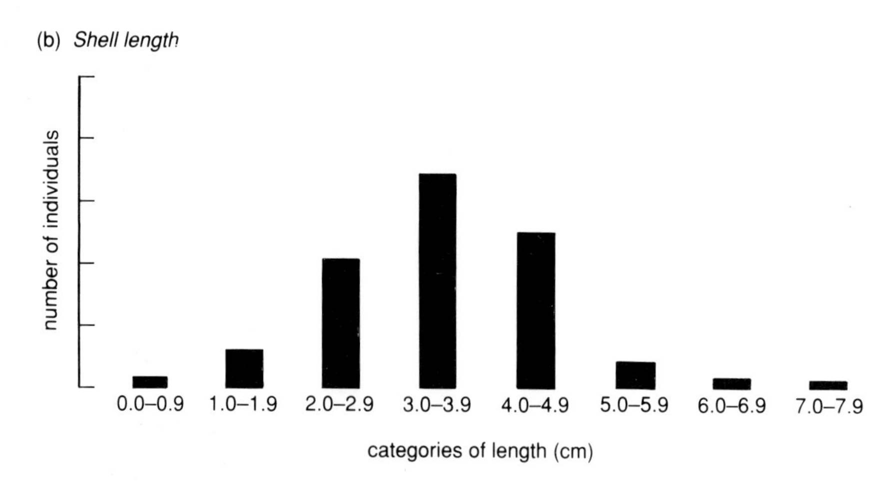

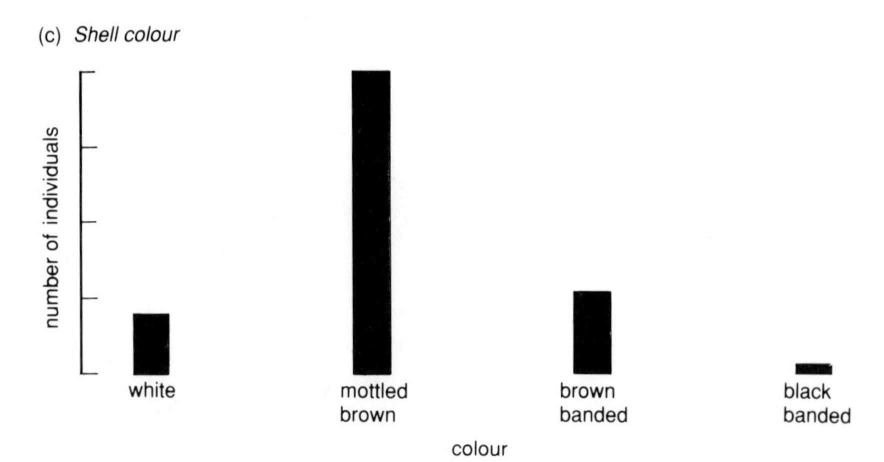

Figure 2 Variation in dog whelk shells

Plant structure and life history

investigation 8

The study of any plant species begins with a description of its structure and life history. A widely distributed species is the Scotch heather, heath or ling (*Calluna vulgaris*), a flowering plant belonging to the family Ericaceae.

C. vulgaris is a low and much-branched evergreen shrub, with minute leaves. It is found on dry acid soils and on wet heathlands and moors. The natural shape of the plant is hemispherical (Figure 1), but the growth form depends much upon environmental factors such as drought, light intensity and wind action. Plants subjected to full sunlight are thick-set and cushion-like. This is also the shape of plants subjected to heavy grazing by animals.

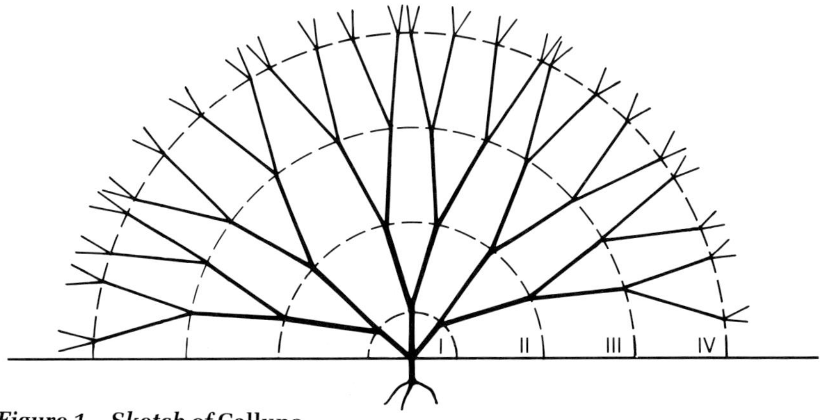

Figure 1 Sketch of Calluna

The side shoots and the leaves are of two different types (Figure 2):

(i) **Long shoots** bearing relatively large, widely spaced triangular leaves, with pointed spurs. These leaves remain active for one season only.

(ii) **Short shoots** bearing smaller, more rounded leaves, which remain active for up to three seasons.

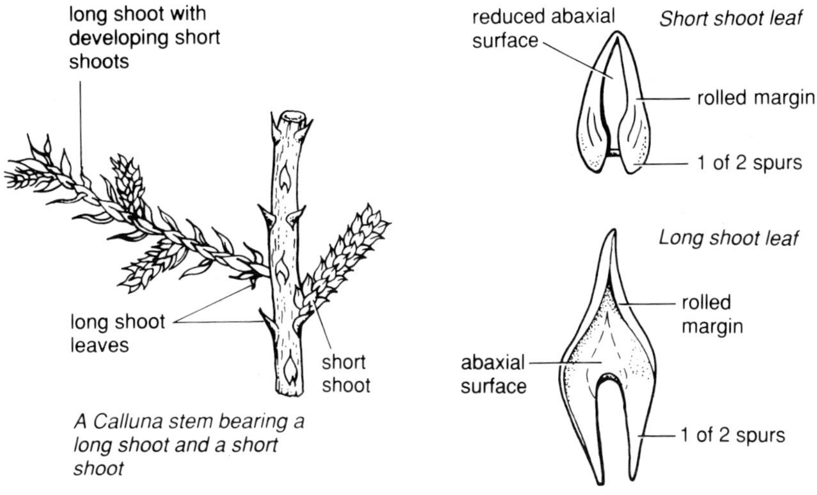

Figure 2 Stem and leaf morphology

Four successive stages in the life cycle of *C. vulgaris* are recognised (Figure 3):

(i) **Pioneer**: plants aged 3–6 years. Annual growth zones are clearly demarcated (Figure 4). Open spaces (e.g. bare earth) occur between individual plants in stands. Sparse flowering.

(ii) **Building**: plants aged 10–15 years, hemispherical, with maximum cover and density of the leaf canopy. No spaces between individual plants. Free flowering.

(iii) **Mature**: plants aged 15–25 years. The woody stems separate at the centre of plants and may touch the ground around the margins. Reduced flowering.

(iv) **Degenerate**: plants aged 25 years or more. Central branches die, leaving a gap. Peripheral branches may root adventitiously. Sparse flowering.

(i) *Pioneer*

(ii) *Building*

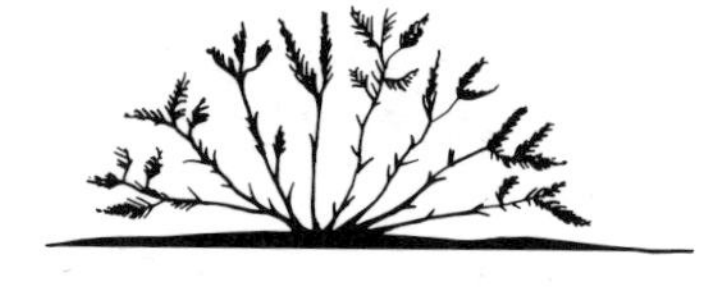

(iii) *Mature*

(iv) *Degenerate*

***Figure 3** Stages in the life history of* Calluna

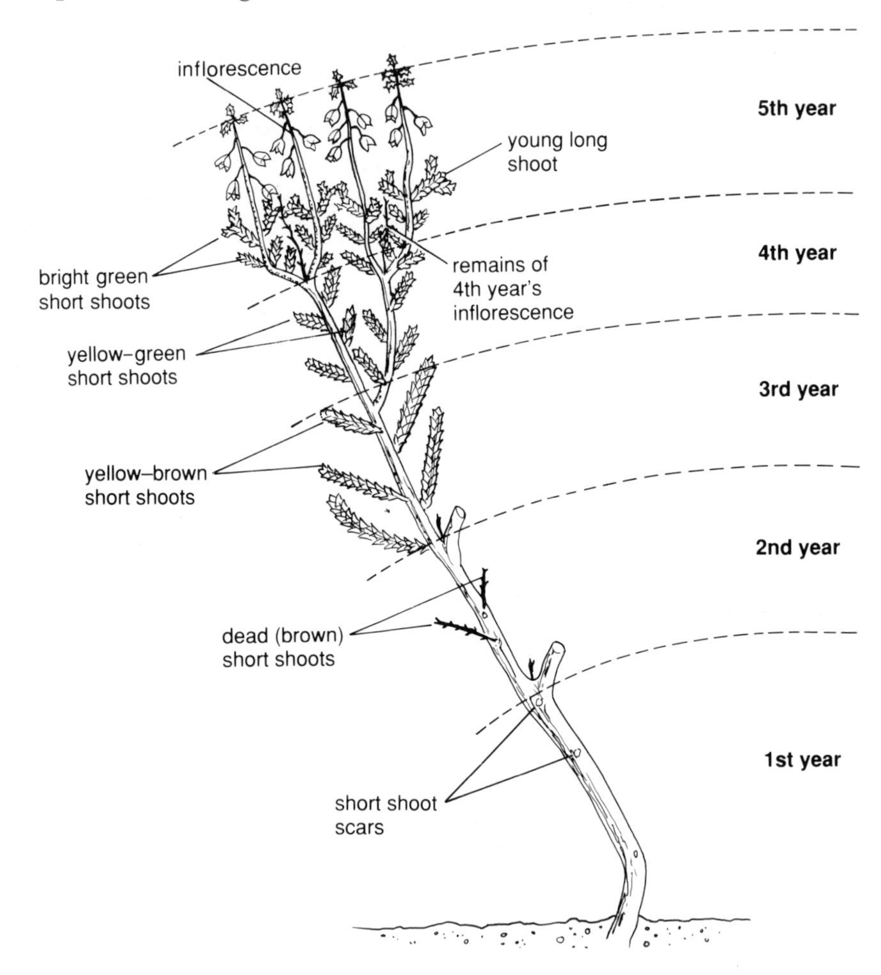

***Figure 4** A five-year-old stem of* Calluna *showing 'storeyed' growth. Short shoots remain on the plant for 2–3 years, then fall leaving a scar*

The purpose of this investigation is to illustrate the structure and life cycle of *Calluna vulgaris* from material collected in the field.

Equipment and materials

- Ordnance Survey map
- pocket knife
- hand lens
- polythene bags

PLANT ECOLOGY

Experimental procedure

1 Visit a heathland area where there are several separate stands of *C. vulgaris*. Determine the age of the plants in these stands by (a) counting the annual growth zones on stems, and (b) counting the number of annual growth rings on stems cut at ground level.
2 Attempt to determine the distribution of plants in the (a) pioneer, (b) building, (c) mature and (d) degenerate stages. Record the distribution of these plants on the Ordnance Survey map.
3 Collect stems of plants aged 3–7 years. Put these into polythene bags and take them back to the laboratory. In the laboratory make accurate scale drawings to show (a) the general structure of a shoot and (b) the differences between long and short shoots.

Taking it further

1 Carry out similar surveys of other heathland species, notably the cross-leaved heather (*Erica tetralix*), bell heather (*Erica cinerea*) and bilberry (*Vaccinium myrtillis*).
2 Obtain cut sections of leaves from the long and short shoots of *C. vulgaris*. Make annotated diagrams to illustrate the anatomical similarities and differences between these two leaves. How are the leaves of *C. vulgaris* adapted to conserve water?
3 Investigate the effects of light intensity on the growth and form of *C. vulgaris* in exposed and shaded sites. Measure light intensity at 1 m above the ground throughout the year. Attempt to relate differences in morphology and anatomy to the amount of light plants have received. Use a potometer to determine rates of transpiration from twigs of uniform mass, cut at different sites. What relationship, if any, exists between the light intensity received by a *Calluna* plant and its rate of transpiration?

Bibliography

Beijerinck, W (1940) *Calluna*: a monograph on the Scotch heather *Verh. Adak. Wet. Amst.* (*3rd section*) 38 1–180.
Watt, A S (1955) Bracken versus heather, a study in plant sociology *J. Ecol.* (43) 490–506.

investigation 9

Estimating plant biomass

The **biomass** of a plant is its total wet or dry mass per unit area of its habitat. Stands of *Calluna vulgaris*, from the pioneer to mature stages, growing on heathlands or moors, provide suitable and accessible material for biomass estimations. From the mass of wet or dried plant material, harvested from plants of different ages, it is possible to show how the total plant biomass of *Calluna* changes throughout its life history. Furthermore, the investigation can be extended by subdividing harvested plant material into its component parts (leaves, stems, roots and flowers) and determining the biomass of each component.

Equipment and materials

- 4 × ½ metre quadrats
- secateurs
- hand fork
- 20 × plastic sacks (with string and marking pen)
- pen knife
- hand lens
- large tins (e.g. bases of biscuit, cake or sweet tins)
- top pan balance
- steam oven

Experimental procedure

1 Select a site with pioneer, building and mature stands of *C. vulgaris*. Determine the age of stands by ring counts at the base of cut stems.
2 Place the four quadrats at selected sites. Use secateurs to cut off all the shoots at ground level, within the area bounded by the quadrats. Transfer the shoots to a plastic sack. Tie the sack and label it with the age of the plants.
3 Use the fork to dig up all the roots within the area bounded by the quadrats. Transfer them to a plastic sack, tie the sack and label it.
4 On returning to the laboratory, determine the mass of the plant material gathered at each site. Record the fresh biomass of each sample.
5 Open the sacks. Transfer the plant material from each sack to a labelled tin. Place the tins into a steam oven and heat them until there is no further change in the mass of the plant material. Record the dry biomass of each sample.
6 Plot a graph to show the relationship between the biomass of fresh and dried samples (see Figure 1).
7 Subdivide the dried shoots into (a) leaves, (b) flowers and (c) stems. Weigh each sample and record its mass. In addition, weigh and record the mass of the roots.
8 Plot graphs similar to Figures 2 and 3 to show how the biomass of each component is related to the age of the plants.

Taking it further

1 Devise and use methods for estimating the biomass of
(a) annual herbaceous plants
(b) perennial shrubs or trees.
How could leaf litter be used to provide estimates of annual leaf biomass?

2 Use a lawn mower, with its blades set at a fixed height above ground level, to make weekly mowings of a unit area of a lawn. Retain the mowings, weigh them, and record their mass. Plot a graph to show how the fresh biomass of harvested mowings varies throughout the year.

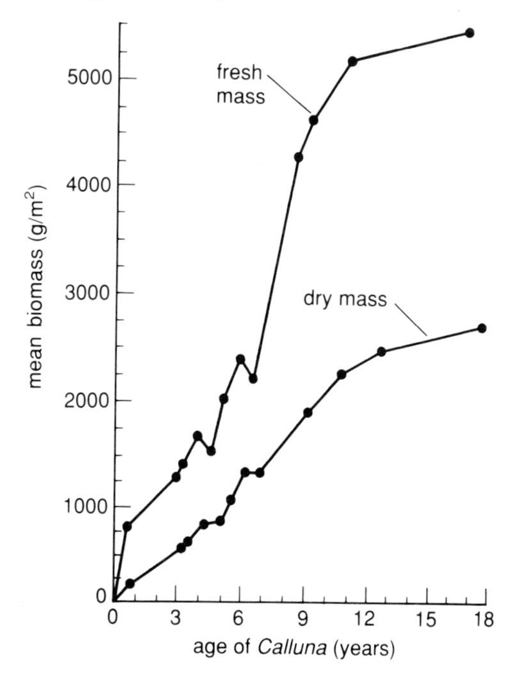

Figure 1 Mean biomass of crop as fresh and dry mass per m² plotted against the age of Calluna *stands (Ashdown Forest, November 1990)*

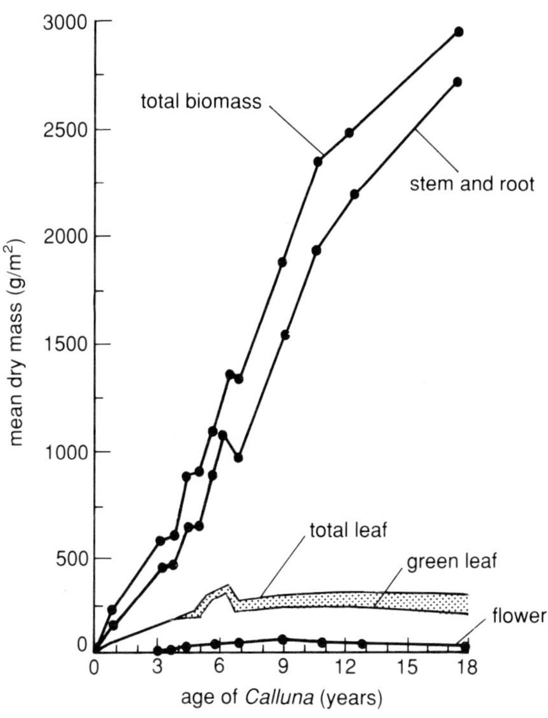

Figure 2 The relationship between the age of Calluna *and the dry mass of its component parts*

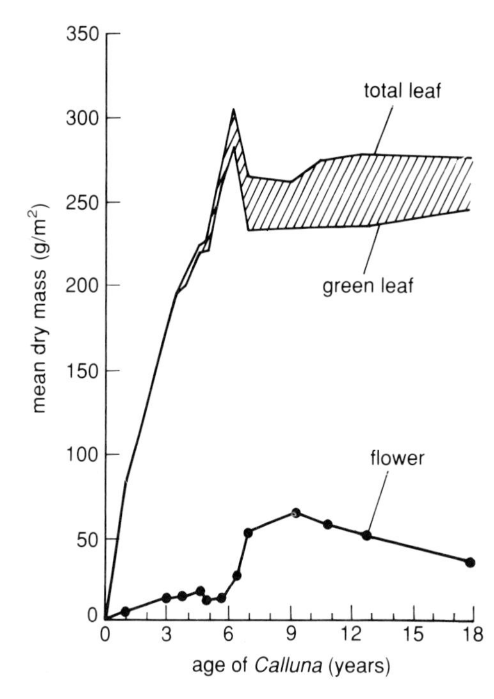

Figure 3 The relationship between the age of Calluna *and the dry mass of the leaves and flowers*

Sun and shade leaves

The leaves of many woodland trees which grow both in full sunlight and dense shade show adaptations according to the amount of light they receive. Typically, the leaves of '**sun**' plants, growing in full sunlight, differ in morphology and anatomy from those of '**shade**' plants which live in full or dappled shade. There are also physiological differences between the two types of leaves, notably in their rates of photosynthesis at any given light intensity. In this investigation you will examine some aspects of the physiological ecology of leafy twigs from sun and shade plants.

Equipment and materials

- 2 × beech twigs, each with 4–5 'sun' leaves
- 2 × beech twigs, each with 4–5 'shade' leaves
- 250 cm^3 glass bottle, fitted with a rubber bung and capillary tubing (see Figure 1a)
- 2 × paper clips
- bench lamp, fitted with a 100 W bulb
- top pan balance
- scissors
- ruler, graduated in mm
- cm^2 graph paper, glued to thick cardboard
- tissue papers

PLANT ECOLOGY

Experimental procedure

Relative rates of photosynthesis

1 Fill the 250 cm^3 bottle with tap water. Insert one of the 'sun' leaves to the position shown in Figure 1a, and replace the bung. Adjust the bung, with its attached capillary tube, so that the meniscus is in the horizontal part of the tube. Set up the apparatus at about 20 cm from an illuminated 100 W electric light bulb, and record the distance travelled by the meniscus in 5 or 10 minutes. Use the formula $\pi r^2 h$ to calculate the volume of gas (oxygen) evolved in unit time.

2 Repeat procedure **1** with one of the 'shade' leaves. Record your results.

Specific leaf areas

3 Take the 'sun' leaf and dry it on tissue paper. Trace its outline onto the mounted graph paper. Calculate the approximate surface area of the leaf by adding together whole (1.0 cm^2), three quarter (0.75 cm^2), half (0.5 cm^2) and quarter (0.25 cm^2) squares (see Figure 1b). Record the approximate surface area of the leaf.

4 Repeat procedure **3** with the 'shade' leaf. Record your results.

5 Cut out a 100 cm^2 square from the mounted graph paper and weigh it. Record its mass. Cut out the leaf tracings, weigh each one and record its mass. Calculate the surface area of each leaf from the following equation.

$$\text{surface area (cm}^2\text{)} = \frac{\text{mass of leaf cut out (g)} \times 100}{\text{mass of 100 cm}^2 \text{ mounted graph paper (g)}}$$

6 Calculate the specific leaf area (SLA) for each leaf:

$$\text{specific leaf area (SLA)} = \frac{\text{leaf area (cm}^2\text{)}}{\text{leaf mass (g)}}$$

7 Devise and use your own method for calculating gas (oxygen) production per unit area of leaf surface in a given time. Show all your working and record your results. What conclusions can be drawn?

Relative rates of transpiration

8 Take a 'sun' leaf and a 'shade' leaf and support each one in a paper clip, as shown in Figure 1c.
Devise and use your own method for calculating water loss per unit area of leaf surface in a given time. Record all your results and write a brief explanation.

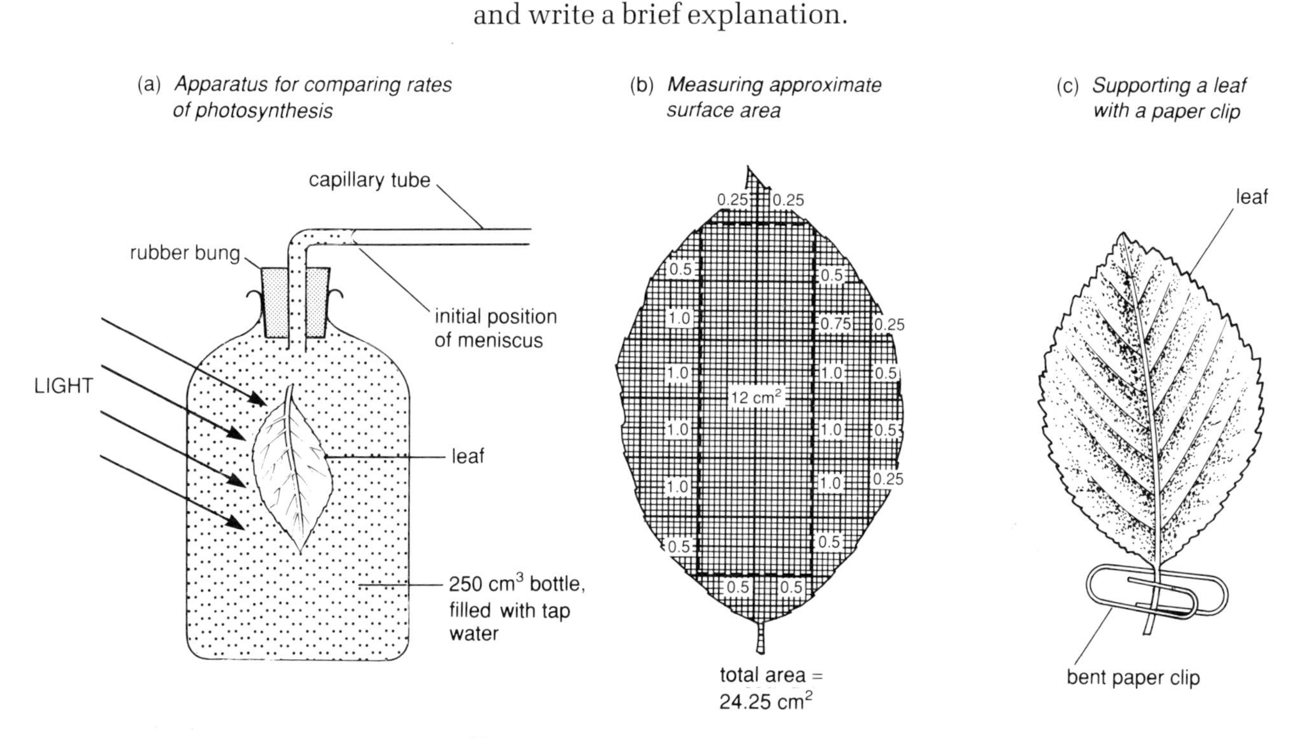

Figure 1 Sun and shade leaves

Taking it further

1 Compare the specific leaf areas, rates of photosynthesis and rates of water loss in 'sun' and 'shade' leaves of (a) hazel and (b) silver birch.
2 Compare the number and distribution of stomata in 'sun' and 'shade' leaves. Paint the leaf surface with clear nail varnish, peel off the dried varnish layer, and examine the imprint under a microscope. Devise and use your own method for determining the number of stomata per unit area of leaf surface.
Examine prepared slides of 'sun' and 'shade' leaves under a microscope. Make drawings to illustrate their anatomical similarities and differences.

investigation 11

Investigating the compensation point

Plants respire and photosynthesise. During respiration they release carbon dioxide; during photosynthesis they take it up. The point at which carbon dioxide production and uptake by a plant reach an equilibrium, with no net exchange between the plant and the atmosphere, is called the **compensation point**. This point differs in sun and shade leaves, as well as in plants that normally grow in full sun, partial shade or complete shade. Relative compensation points may be determined by using leaf discs suspended above a CO_2-sensitive colour indicator, hydrogen carbonate solution. The colour changes in this indicator, and the processes that cause these changes, are shown below.

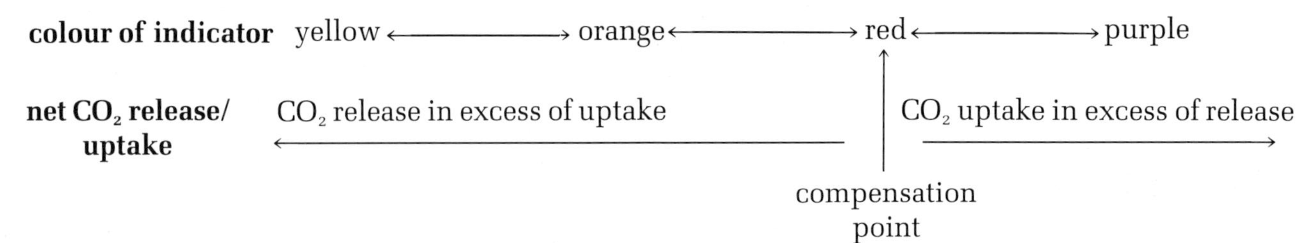

Equipment and materials

- sun leaf of beech
- shade leaf of beech
- leaf of bramble
- 4 × boiling tubes, fitted with rubber bungs
- test tube rack
- hydrogen carbonate solution
- 100 cm^3 beaker
- glass tube
- number 12 cork borer
- 5 cm^3 plastic syringe
- sewing needle
- cotton
- ruler
- bench lamp, fitted with a 60 W electric light bulb
- clock
- glass-marking pen

Experimental procedure

1 Prepare 50 cm^3 hydrogen carbonate indicator solution according to the manufacturer's instructions. (This usually involves making a ten-fold dilution of indicator concentrate.) Pour the diluted solution into the beaker. Insert the glass tube into the beaker and gently blow through it until the indicator is red. Withdraw some of this red indicator solution and put it into one of the boiling tubes. Fit an air-tight rubber bung to the tube and use this indicator solution as a colour standard.

2 Continue to blow air into the beaker of indicator solution until it is orange-yellow.
3 Take the remaining boiling tubes and number them from **1–3**. Make ink marks at distances of 1.0 and 2.5 cm from the base of each tube. Transfer the tubes to the test tube rack. Use the syringe to transfer yellow-orange indicator solution into each tube to the 1.0 cm mark.
4 Use the cork borer to cut two discs from each leaf. Thread a piece of cotton through the centre of each disc. Transfer two leaf discs to each tube as follows:
tube 1: sun leaf of beech
tube 2: shade leaf of beech
tube 3: leaf of bramble
Suspend the discs as shown in Figure 1, with their lower margins at the 2.5 cm mark. Fit a rubber bung to each tube, ensuring that it is air-tight.
5 Set up the test tube rack at a distance of 15 cm from the illuminated electric light bulb. Record the time taken for the indicator solution in each tube to change colour from yellow-orange to red. Record your results.
6 What is being determined in this investigation? How does it relate to the compensation point of the leaves?

Taking it further

1 Compare changes in the compensation points of beech, oak and holly leaves during the year.
Alternatively, use the apparatus (Figure 1) in the field, together with a light meter, to measure light intensity. Compare the compensation points of leafy herbaceous plants growing on the sunny and shaded sides of walls or hedges.
2 Hydrogen carbonate indicator solution can be used to determine the relative rates of respiration in invertebrates. Devise and use apparatus for comparing the rates of respiration in earthworms, woodlice and slugs.

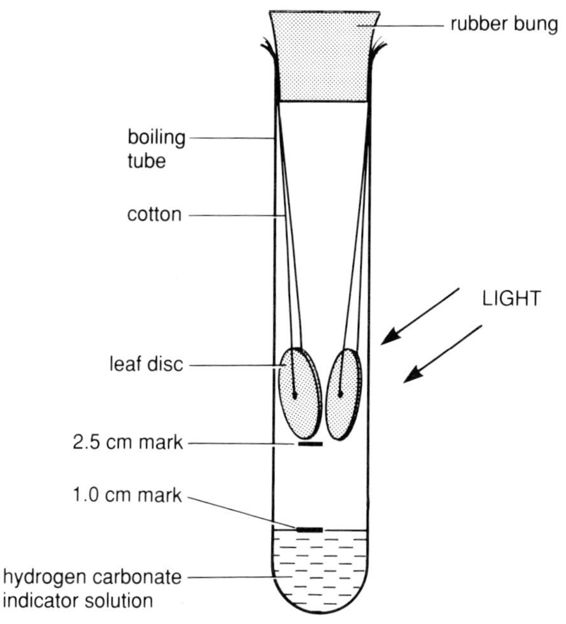

Figure 1 Apparatus for finding the compensation point of leaves

investigation 12

Competition between duckweeds

Plants compete with one another for space. One of the most convenient groups of plants for laboratory studies of competition between species is duckweeds. These are small floating pond plants, abundant in summer, with numerous leaf-like structures called platelets or thalli. Each platelet or thallus has some of the properties of both leaves and stems. There are five different species of duckweed found in the UK (Figure 1). These are common duckweed (*Lemna minor*), gibbous duckweed (*Lemna gibba*), ivy duckweed (*Lemna trisulca*), great duckweed (*Lemna polyrhiza*) and rootless duckweed (*Wolffia arrhiza*). All of these plants are free-floating at the surface of the water, with the exception of *L. trisulca*, a plant that is normally completely submerged at a depth of 1–5 cm below the water surface.

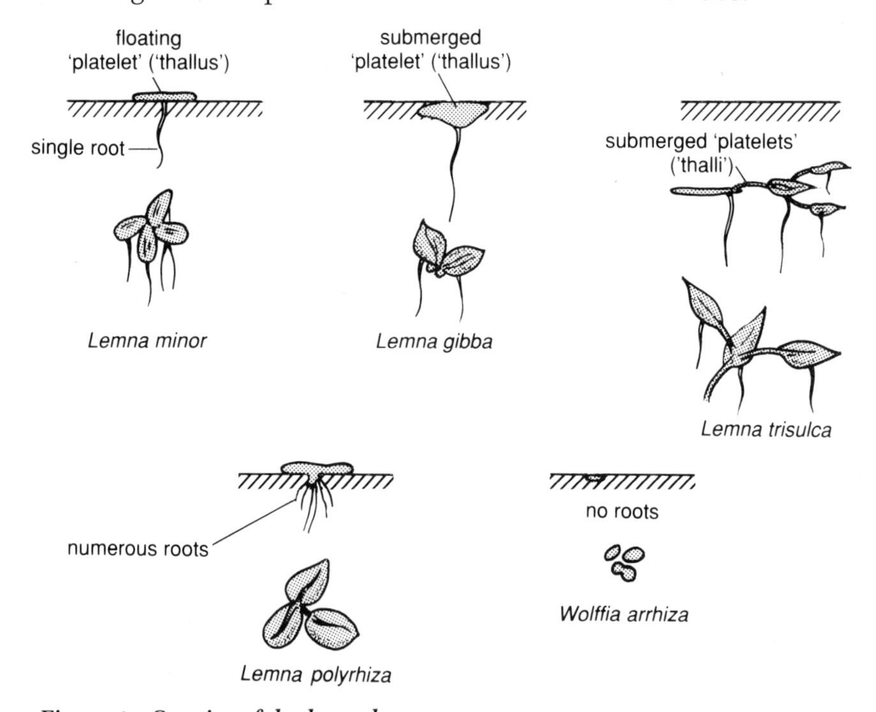

Figure 1 Species of duckweed

In this investigation you will compare the rate at which populations of a duckweed increase both (a) in isolation and (b) in competition with another species.

Equipment and materials

- *Lemna minor* and *Lemna trisulca*
- 3 × 1 dm^3 beakers
- 2.4 dm^3 Sach's complete culture solution
- distilled water
- 250 cm^3 measuring cylinder
- bench lamp, fitted with a 100 W or 150 W electric light bulb
- clingfilm
- glass-marking pen

Experimental procedure

1 Prepare 2.4 dm^3 Sach's complete culture solution according to the manufacturer's instructions.
2 Number the beakers from 1 to 3. Using the measuring cylinder, pour 800 cm^3 Sach's complete culture solution into each beaker. Mark the level of the solution on the outside of each beaker.
3 Place duckweed into the beakers as indicated below:
beaker 1: 6 platelets or thalli of *Lemna minor*
beaker 2: 6 platelets or thalli of *Lemna trisulca*
beaker 3: 3 platelets or thalli of *Lemna minor*, and 3 of *Lemna trisulca*
Cover each beaker with clingfilm to reduce water loss.
4 Set up the electric lamp above the beakers and arrange them so that they are fully illuminated. (Alternatively, set up the beakers in a greenhouse during the period April–September.) Add distilled water to the beakers to replace water lost by evaporation. The experimental apparatus is shown in Figure 2.
5 At weekly intervals, make counts of the numbers of platelets (thalli) of each species. Plot graphs to show changes in population density when the two species of duckweed are (a) grown in isolation and (b) grown in competition with one another. Try to explain your results.

Taking it further

1 Select one or more species of duckweed and place it (them) in buffered solutions of mineral salts, over a pH range of 3–10. Try to find the pH that is optimal for the growth of this plant. Do other species of duckweed have the same pH requirement for optimal growth?
2 Grow one or more species of duckweed in competition with the water fern *(Azolla filiculoides)*. Plot graphs to show the course and outcome of competition between the two plants.

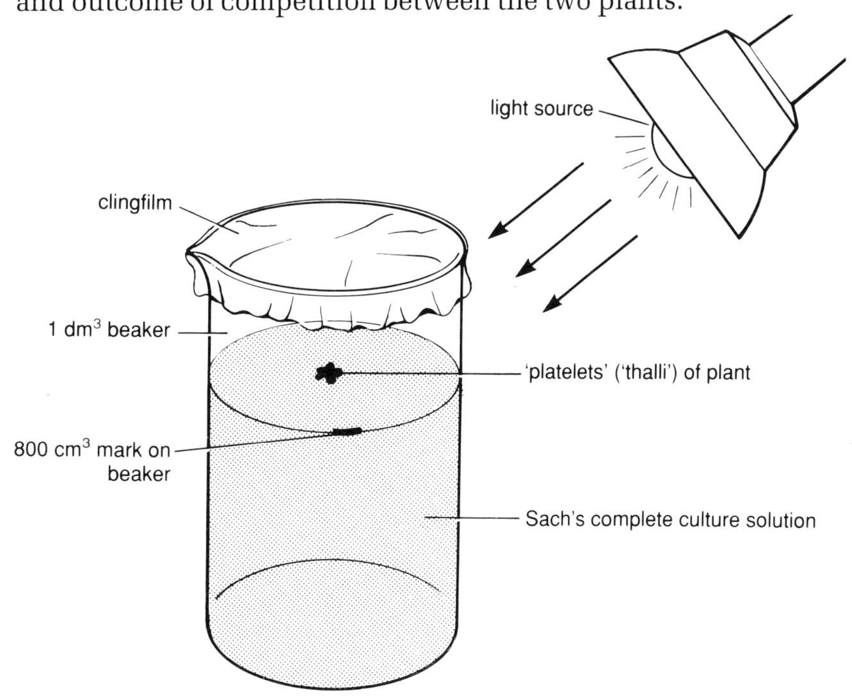

Figure 2 Apparatus for growing duckweeds

13 *investigation*

Investigating allelopathy 1: resins and tannins in leaves

The leaves of many trees contain nutrients that provide food for primary consumers, notably small invertebrates such as insects. Generally, however, these herbivores do not wreak havoc on their food plants because the plants have evolved some protective devices which act as herbivore deterrents. Among these is the production of specific protease inhibitors, such as resins in conifers, and tannins in broad-leaved trees. In this investigation you will find out if the leaves of common and widely distributed trees produce protease inhibitors capable of interfering with the digestive processes of those animals that feed on them.

Equipment and materials

- 3 × leaves of conifers (e.g. Scot's pine, larch, cupressus)
- 3 × leaves of broad-leaved trees (e.g. oak, beech, alder, hornbeam, sycamore, cherry, holly)
- 2 × 'Marvel' milk agar plates
- 5 cm^3 1% $^{w/v}$ trypsin solution
- distilled water
- pestle and mortar
- 2 × 2 cm^3 plastic syringes
- number 6 cork border
- ruler
- incubator, maintained at 25° C
- glass-marking pen
- knife or scalpel

Preparation

Prepare the Marvel milk agar plates as follows: add 1 g Marvel milk and 1 g bacteriological agar powder to 50 cm^3 distilled water. Transfer the mixture to a screw-top glass bottle and autoclave. While the agar is still molten, pour 25 cm^3 into each of two petri dishes. Allow the agar to cool and harden.

Experimental procedure

1 Take the two petri dishes containing the milk agar and number them **1** and **2**. Use the cork borer to cut four wells in each agar plate, arranged as shown in Figure 1. Number the wells 1 to 4 on the bottom of each dish.
2 Cut the leaves of the conifers into 1 cm lengths. Use the cork borer to cut discs from the leaves of the broad-leaved trees.
3 Put three conifer leaf segments into the mortar, add 2 cm^3 distilled water and grind the leaves to a slurry. Transfer two drops of the leaf extract to well 2 of dish 1. Wash out the mortar.
4 Repeat procedure **3** with the second conifer leaf. Transfer two drops of the leaf extract to well 3 of dish 1. Again, repeat this procedure with the third conifer leaf, transferring the extract to well 4.

5 Transfer two drops of trypsin solution to each well of dish 1. Add two drops of distilled water to well 1.
6 Replace the lid of the dish and incubate it at 25°C until translucent zones appear around each of the wells. Measure the diameter of each translucent zone. Use the formula πr^2 to calculate the area of each translucent zone. Calculate the percentage inhibition caused by each leaf extract.
7 Repeat procedures **3–7** with leaf discs from three broad-leaved trees. In each case grind two leaf discs with 2 cm^3 distilled water.
8 Present your results in the form of a histogram. List the trees in rank order, according to their ability to inhibit trypsin activity.

Taking it further

1 Try to find out if leaves contain (a) amylase and (b) lipase inhibitors. Use a 2%$^{w/v}$ starch agar for your investigation of amylase inhibition, and a 2%$^{w/v}$ trybutyrin or mayonnaise agar for your investigation of lipase inhibition.
2 Devise and carry out an investigation to find out how tannin levels in leaves change during the growing season.
Estimate the tannin content of the leaves by a colorimetric method. Prepare a 0.1%$^{w/v}$ ferric sulphate solution, and tannic acid solutions ranging from 0.01 to 0.1%$^{w/v}$. Mix the tannic acid solution (2 parts) and ferric sulphate solution (1 part) to obtain a blue-green coloration. Determine the absorbance of these mixtures over the range of tannic acid concentrations. Plot a standard curve and use this curve to estimate the concentrations of tannic acid in the leaves of broad-leaved trees.

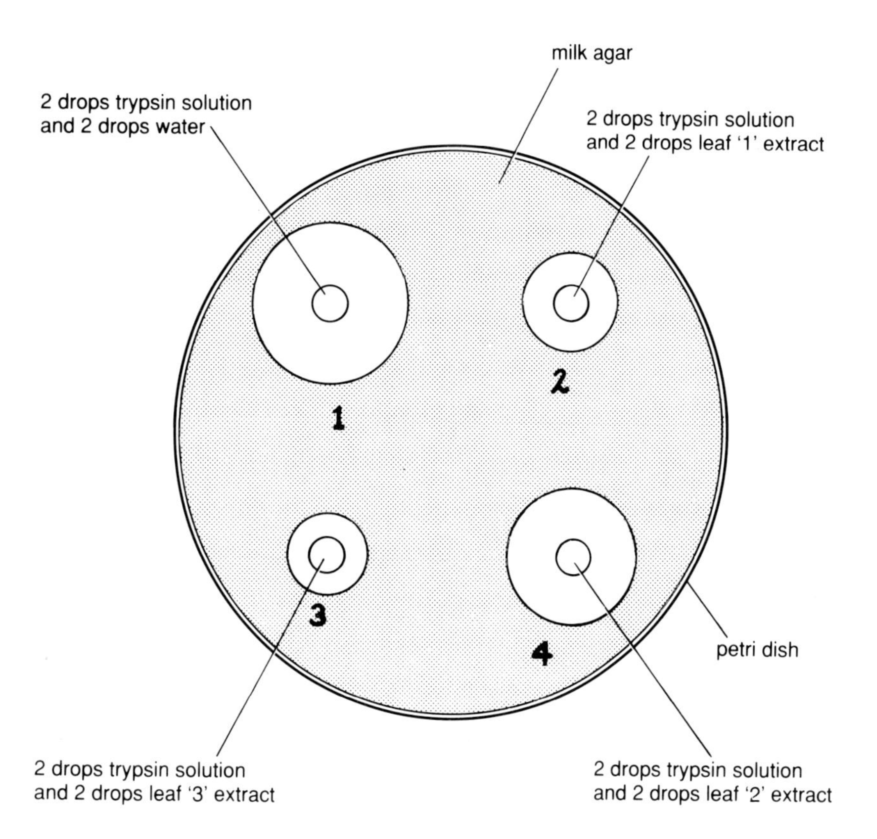

Figure 1 Arrangement of the wells in a milk agar plate

investigation 14

Investigating allelopathy 2: seed germination

We saw in Investigation 13 how resins and tannins from leaves inhibit proteolytic enzymes, thereby protecting the plants that contain them from grazing animals. In the wild, these and other allelopathic substances perform a useful ecological function by inhibiting the growth of rival plant species. This gives the plants that produce them a competitive advantage over their rivals, mainly by preventing colonisation of adjacent bare soil by competitors. Allelopathic substances are produced by many different plants, notably in their leaves (walnut, sunflower, heather, rhododendron, holly etc.) or roots (barley, grass, oak). In this investigation, the presence of allelopathic substances is detected by their ability to inhibit the germination of certain seeds such as mustard, cress, lettuce or grass.

Equipment and materials

- leaves of rhododendron, yew, holly, pine or heather (freshly picked)
- 200 mustard seeds
- 250 cm^3 distilled water
- 100 cm^3 measuring cylinder
- 2 × 2 cm^3 plastic syringes
- 2 × 9 cm petri dishes
- filter funnel
- filter papers
- scissors
- forceps
- kitchen mixer
- top pan balance
- glass-marking pen

Experimental procedure

The procedure is shown in Figure 1.

1 Weigh out 50 g leaves. Transfer the leaves to the kitchen mixer, add 200 cm^3 distilled water, and grind the leaves to a slurry.

2 Place the filter funnel over the measuring cylinder. Line the filter funnel with filter paper, and filter the slurry. Retain the filtrate in the measuring cylinder.

3 Line the two petri dishes with filter paper (cut to size with scissors).

4 Add 2 cm^3 distilled water to one of the dishes and 2 cm^3 leaf filtrate to the other. Label the dishes.

5 Use forceps to sow 100 mustard seeds in each dish, spacing the seeds so that they do not touch one another, or the sides of the dishes.

6 Replace the lids of the dishes. Place the dishes on a flat surface. Observe the seeds over a period of 3–10 days.

7 For each dish record (a) the number of seeds that have germinated, and (b) the time taken for 50% of the seeds to germinate. What do you conclude from your results?

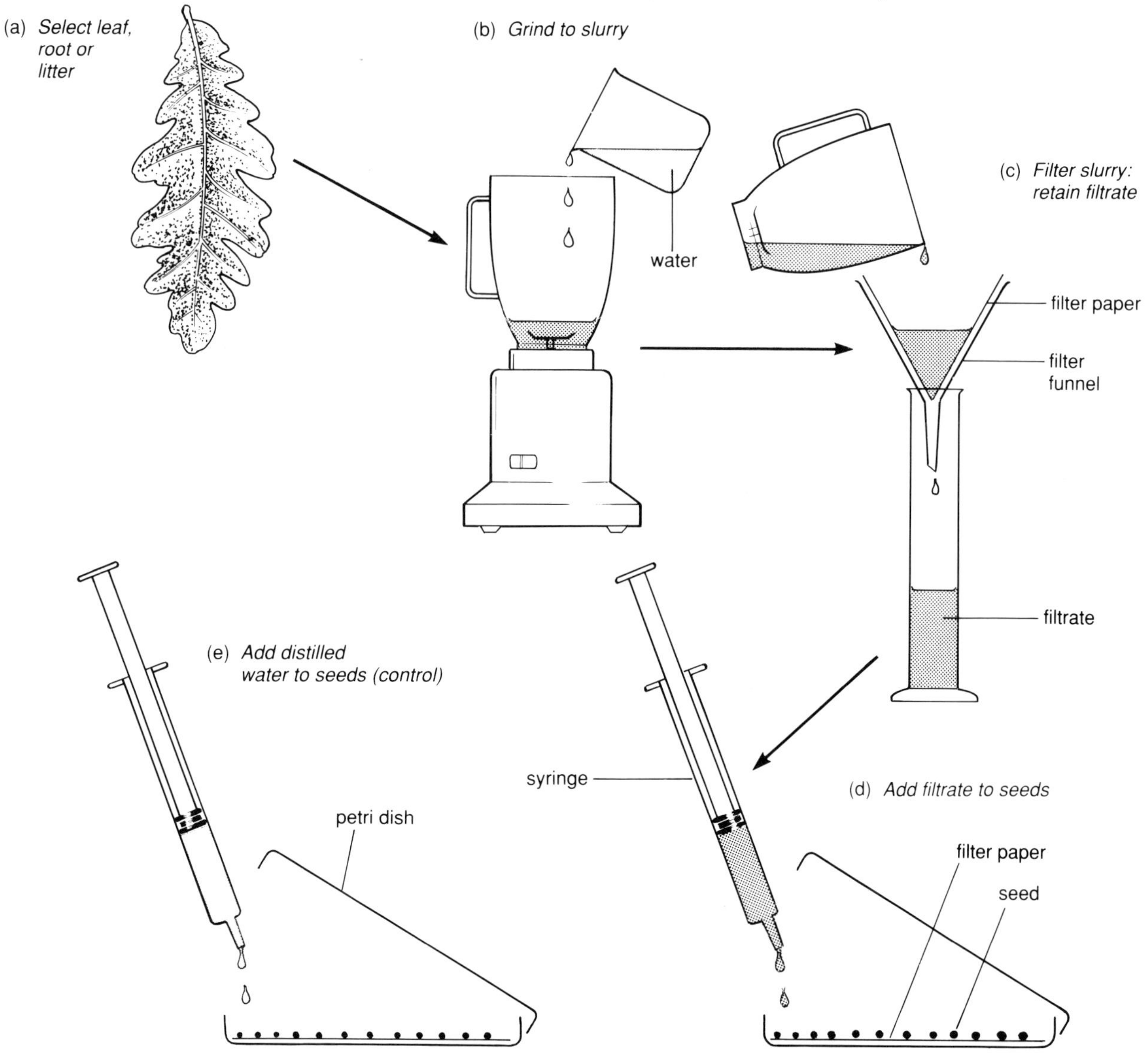

Figure 1 ***Stages in testing for the presence of an allelopathic substance***

Taking it further

1 Repeat the investigation using seeds of (a) lettuce, (b) cress and (c) grass. How do your results compare with those obtained with mustard seeds?

2 Design and carry out an investigation to find out how the amounts of allelopathic substances produced by leaves vary throughout the growing season. Extend your investigation to include leaf litter in various stages of decomposition.

3 Select a herbaceous plant community such as grassland, heathland, or the herb layer of a wood. Try to find out if any plants in these communities produce allelopathic substances either from (a) their leaves or (b) their roots, which inhibit (i) seed germination or (ii) growth of other species in the same community.

investigation 15

Nitrogen fixation in plants

Certain plants found in fresh water, notably the water fern *Azolla* and the cyanobacterium *Anabaena*, can fix atmospheric nitrogen as nitrate (NO_3^-). Bacteria and other organisms in the water may then reduce this compound to nitrite (NO_2^-), a substance that is highly toxic to some aquatic animals.

Water fleas (*Daphnia*) are one of a number of common aquatic animals that are sensitive to levels of nitrite (NO_2^-) in the water. In this investigation, nitrogen-fixing organisms are added to population of water fleas and their efforts on population density monitored over a period of several weeks or months.

Equipment and materials

- 150 water fleas (*Daphnia*)
- water fern (*Azolla*)
- *Anabaena* culture
- dried yeast
- 2.4 dm^3 Sach's nitrogen-deficient water culture solution
- distilled water
- Merckoquant nitrate-nitrite-sensitive reagent sticks
- 250 cm^3 measuring cylinder
- 20 cm^3 plastic syringe
- 3×1 dm^3 beakers
- clingfilm
- elastic bands
- bench lamp, fitted with a 100 or 150 W electric light bulb
- glass-marking pen

Experimental procedure

1 Prepare 2.4 dm^3 Sach's nitrogen-deficient culture solution according to the manufacturer's instructions.
2 Number the beakers from 1 to 3. Using the measuring cylinder, pour 880 cm^3 Sach's nitrogen-deficient water culture solution into each beaker. Mark the level of the solution on the outside of each beaker.
3 Put 50 water fleas into each beaker, and add a single granule of dried yeast to provide food.
4 Make additions to the beakers as follows:
beaker 1: water fern
beaker 2: *Anabaena*
beaker 2: control (no further additions)
Cover each beaker with clingfilm to reduce water loss. Set up the electric light bulb above the beakers. Switch on the light and ensure that the contents of the beakers are fully illuminated. (Alternatively, set up the beakers in a greenhouse during the period April–September.) Add distilled water to the beakers to replace any water lost by evaporation. The experimental apparatus is shown in Figure 1.

5 At weekly intervals, dip a Merckoquant nitrate-nitrite-sensitive reagent stick into each beaker. Record the levels of nitrate and nitrite in each solution as estimated by this semi-quantitative method.

6 Every fortnight, make a count of the number of water fleas in each beaker. If it is not possible to make a count by direct observation, withdraw 20 cm^3 from each solution using a syringe. Count the number of water fleas in this sample, then multiply the figure by 40 to give an estimate of the total number of water fleas in the beaker.

7 Plot graphs to show changes in the levels of nitrate and nitrite in each beaker during the course of the investigation. To these graphs add changes in population density of the water fleas. What do you conclude?

Taking it further

1 Devise and set up investigations to find out whether *Azolla* or *Anabaena* is the most efficient at fixing atmospheric nitrogen. What do you conclude?
Attempt to find out if nitrogen fixation by *Azolla* is related to its rate of photosynthesis, or independent of it.

2 Determine the effect of different concentrations of (a) sodium nitrite and (b) calcium nitrite on the survival rate of *Daphnia*. Use concentrations of nitrites within the range 0.001–0.01%$^{w/v}$. Plot graphs to show how these compounds affect the survival rate. What do you conclude?

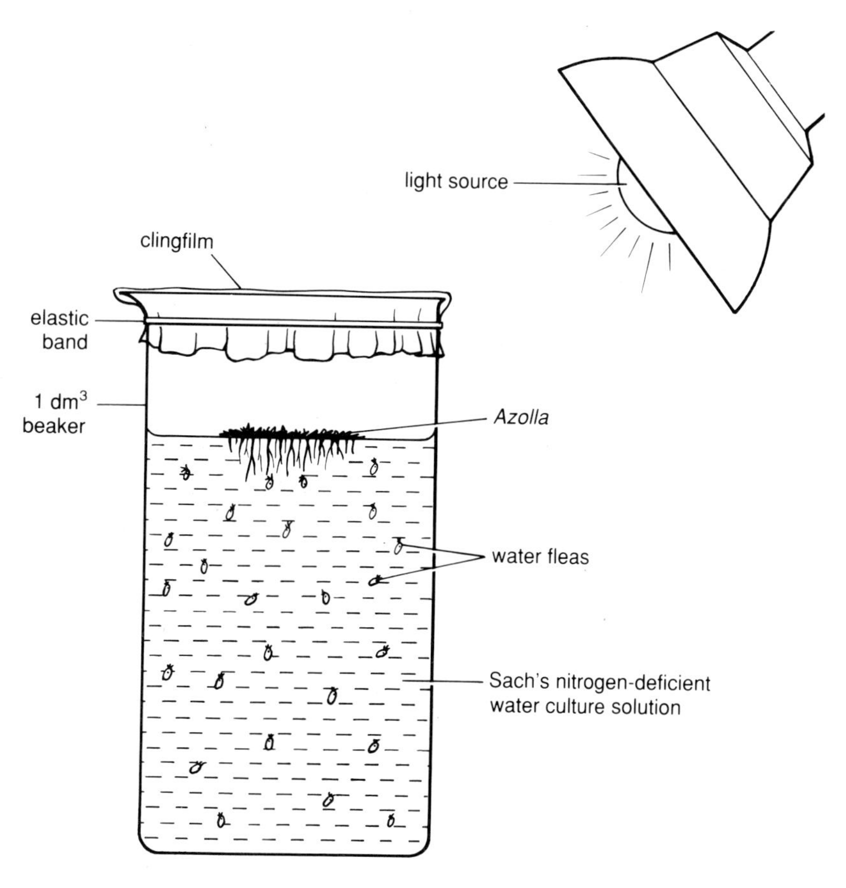

***Figure 1** Apparatus for demonstrating the effects of certain plants on populations of water fleas*

investigation 16

Insects feeding on *Calluna vulgaris*

Insects are the principal herbivores feeding on *Calluna vulgaris*. These primary consumers co-exist by grazing the plant in succession throughout the year. In fact, very little activity is possible at temperatures below 8°C and this alone restricts most insect species to a period of activity between week 16 (April 16) and Week 44 (November 4). Within this period of activity, many insect populations build up their numbers to a maximum which coincides with maximum temperatures in June, July and August. During the late summer and autumn, numbers fall steadily; however, a late warm spell in September or October may cause a temporary rise in population density, or even lead to the reappearance of some adult populations.

Changes in the population density and the succession of insects feeding on *C. vulgaris* may be monitored by making weekly sweepings in heather stands.

Equipment and materials

- keys to the identification of insects
- sweep net
- screw-top jar or polythene bag

Experimental procedure

1 Select a stand of *C. vulgaris* in the age range 5–15 years.
2 Each week, preferably on the same day and at the same time of day, make 100 sweepings while walking through the stand of heather plants. Retain all the insects caught in the sweepings.
3 Return to the laboratory and identify the insects that have been caught. Record the numbers of each species.
4 After sweepings have been taken for one year, or over the period April–October, plot one or more of the following graphs.
  a) Changes in the population density of one or more species (see Figure 1).
  b) The succession of population density peaks shown by primary consumers (see Figure 2).
  c) The duration and succession of inset populations grazing on *C. vulgaris* (see Figure 3).

Taking it further

1 Identify the species that (a) graze on *C. vulgaris* throughout the year and (b) show a seasonal grazing pattern.
Try to find out how numbers of the continuous grazers are affected by temperature, relative humidity and light intensity.
2 Carry out a similar survey on a tree or woody shrub. Place a ground sheet beneath one or more low-lying branches. Strike the branches with a heavy wooden stick or metal bar. Retain those insects that fall onto the ground sheet. Identify these insects and

record their numbers. Plot graphs to show (a) changes in the population density of a named species and (b) successional grazing throughout the year.

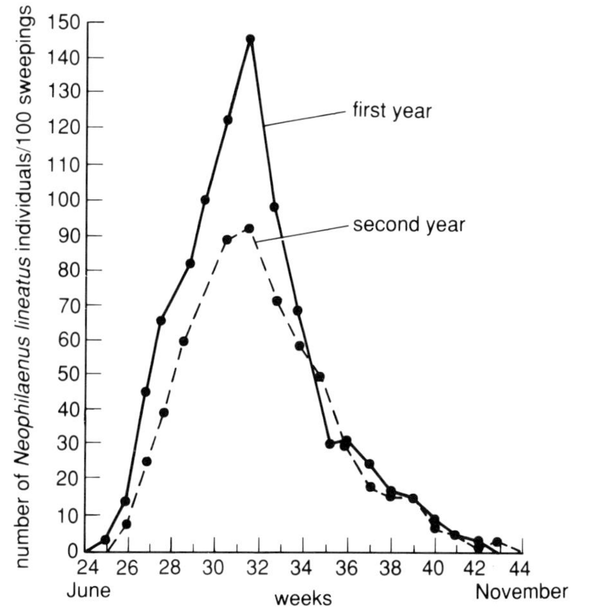

Figure 1 Changes in the population density of **Neophilaenus lineatus** *in successive years*

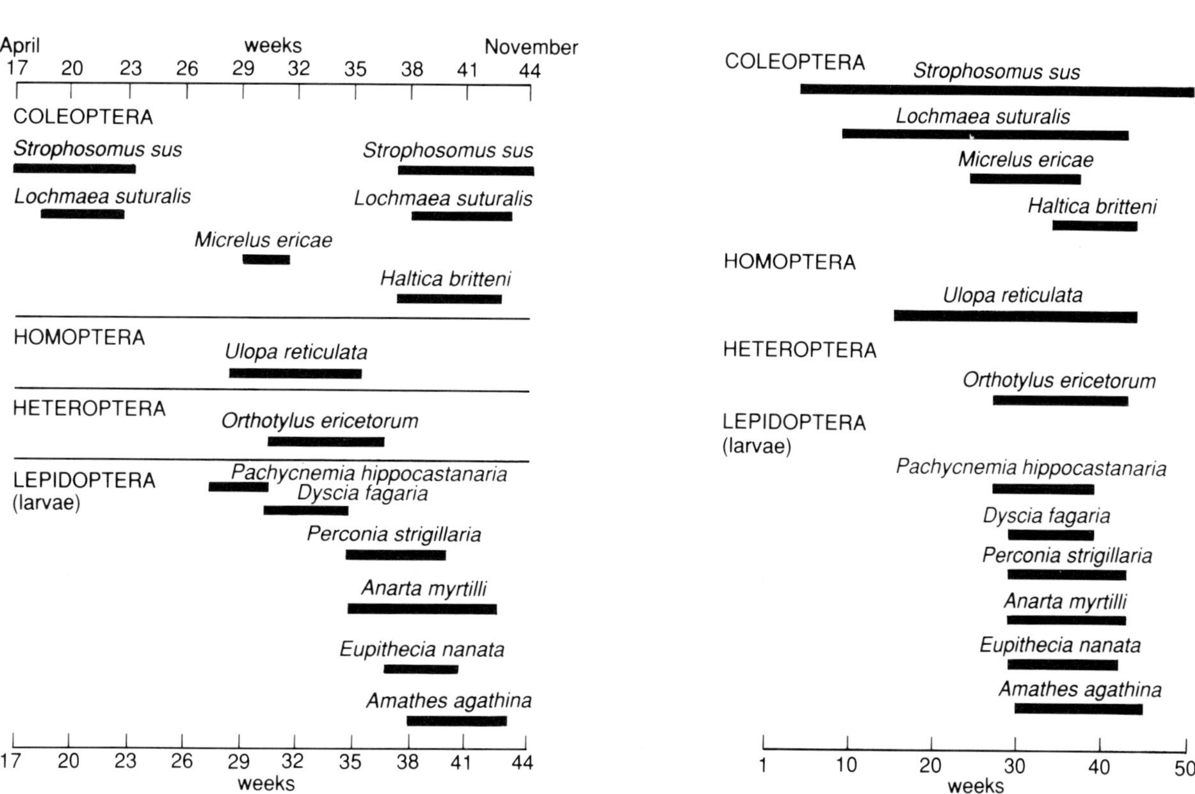

Figure 2 The succession of population density peaks shown by primary consumers on Calluna

Figure 3 The duration and succession of insect populations grazing on Calluna

Bibliography

Keys to the identification of insects Royal Entomological Society, London
Wayside and woodland series Frederick Warne and Co Ltd, London
Invertebrate Types Ginn and Co Ltd, London.

17 investigation

Catching invertebrates in traps

Pitfall traps, set at ground level, catch mostly woodlice, ground beetles, spider, harvestmen, slugs and snails. The aim of this investigation is to compare the numbers of invertebrates caught in baited and unbaited traps. These traps, made from jam jars, tin cans or glass beakers, are laid by digging a hole in the ground so it is deep enough to take the whole container. The rim of each container is then set at ground level (Figure 1) and protected from rain by a tile or wooden board, supported by piles of stones. In this investigation, 32 traps are set up in pairs within a 4 m^2 area marked out by using quadrats. One trap of each pair is baited, either with meat (to attract carnivores) or with succulent nutritious plant materials (to attract herbivores and detritivores). Results are analysed to find out if baiting increases the catch.

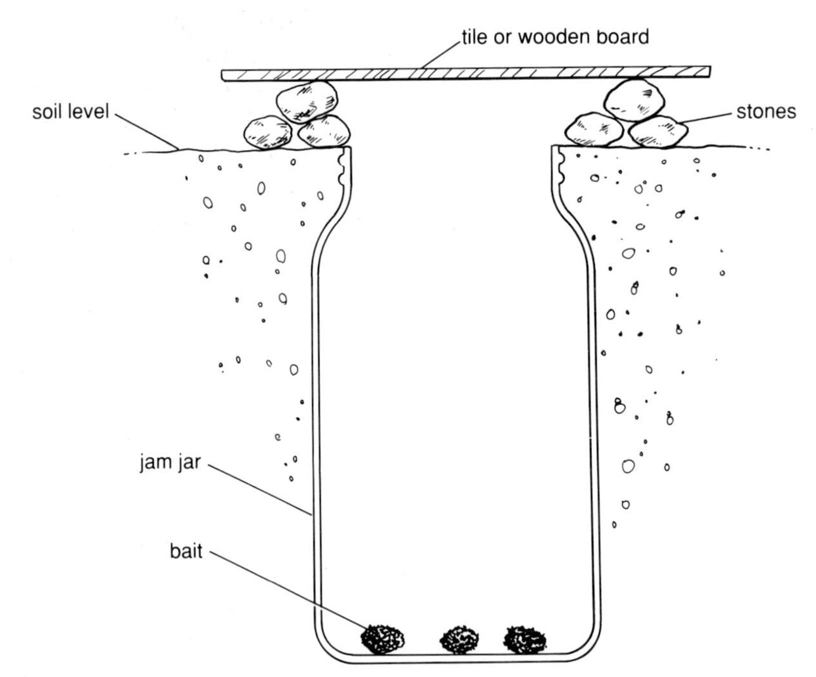

Figure 1 A pitfall trap

Equipment and materials

- 1 m^2 quadrat
- 32 jam jars
- 32 tiles or wooden boards
- garden trowel
- minced beef, diced cold meat or bacon (bait for carnivores)
- diced banana, potato, carrot or apple (bait for herbivores)
- string
- wooden pegs
- hammer
- pebbles
- glass-marking pen

Experimental procedure

1 Use the quadrat, string and pegs to map out an area of 4 m^2, as indicated in Figure 2.

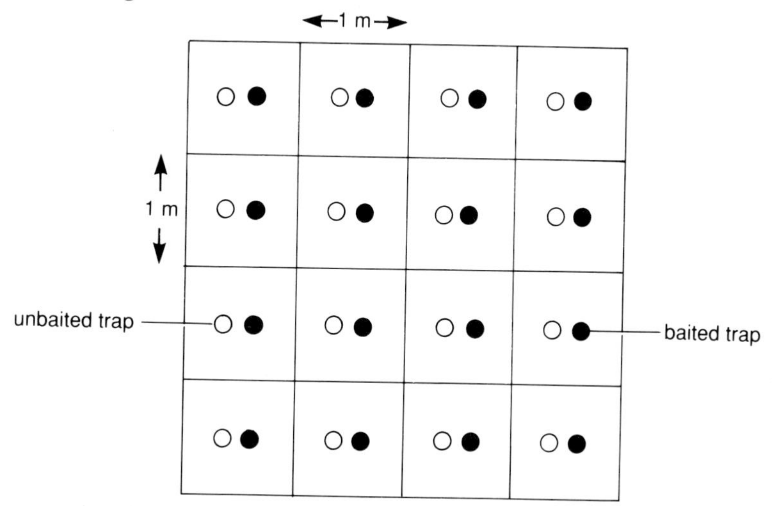

Figure 2 Layout of the pitfall traps

2 Label the jam jars '1–16 unbaited' and '1–16 baited'.
3 At the centre of each 1 m^2 area, dig two adjacent holes, each deep enough to hold a jam jar buried to the level of its rim.
4 Replace the soil around the neck of each jar, flush with the surrounding soil. Place a tile or wooden board above each jar, supported by piles of stones.
5 Bait one half of the traps, as indicated in Figure 1. Use meat as a bait for carnivores (carabid beetles, centipedes, harvestmen, spiders) and fruit or vegetables as bait for herbivores and detritivores (woodlice, slugs, snails).
6 Leave the traps overnight. Next morning, remove them and take them back to the laboratory.
7 Identify the extracted animals from keys such as those derived by Lewis and Taylor (1967) or Bayne, Evans, Llewellyn-Jones and Shalders (1988). Tabulate the numbers of each species caught in the baited and unbaited traps.
8 Examine for significance the difference between samples collected in (a) unbaited and (b) baited traps. Apply, for example, a non-parametric test such as the 'Wilcoxon matched pairs signed ranks' test.

Taking it further

1 Compare the flora and fauna of south-facing and north-facing slopes on grassland or heathland. To what extent does topography affect the flora and fauna of the regions you have studied?
2 From the animals you have captured in traps, construct (a) food chains and (b) food webs of species in a habitat.

Bibliography

Bayne, D, Evans, D, Llewellyn-Jones, J and Shalders, J (1988) *Biokeys* Blackie
Lewis, T and Taylor, L R (1967) *Introduction to Experimental Ecology* Academic Press.

investigation 18

Estimating population density

In some instances it is possible to make fairly accurate estimates of the population density of insect species by directly counting the number of individuals within small sample areas. This method is particularly applicable to populations of large insects, such as grasshoppers and froghoppers, on lawns, in meadows, or in stands of heather. Squares of green or brown cloth, usually 1 m^2, are pegged out on a lawn or in a meadow. Counts are then made of the number of individual animals on the cloth at given periods throughout the day. Similar counts can be made in stands of heather or bracken. Here, it is best to cut a 1 m^2 clearing in the vegetation and to observe insects against a background of bare earth. This simple counting technique can be used to show how the numbers and species of insects change throughout a 12 or 24 hour period.

Equipment and materials

- 1 m^2 green or brown cloth
- secateurs
- pegs
- sweep net
- keys to identification of insects

Experimental procedure

1 Select a suitable site, such as a lawn, meadow, or stand of heather or bracken.
2 On a lawn or in a meadow, spread out the cloth and peg it at the corners. In a stand of heather or bracken, use secateurs to cut a 1 m^2 clearing (Figure 1).
3 Position yourself with a good view of the cloth or clearing. Try to avoid shading the area that is under observation.
4 Starting early in the day, make observations and counts of insects at 30 or 60 minute intervals. If necessary, net insects and identify them from keys.
5 Record the numbers of each species observed on the cloth, or in the clearing, in 10–20 samples. Plot your results in the form of histograms to show (a) how the numbers of each species vary throughout the day and (b) the daily succession of insect species.

Taking it further

1 Using the technique outlined above, record and plot changes in the population of one or more adult insects during the period April–August.
2 Use a sweep net to sweep a stand of heather or gorse 100 times per week throughout the year. Identify the insects you have caught from keys. Plot graphs to show changes in the population

density of one or more species. Try to classify your samples into those animals feeding at each of thc following trophic levels:

1st trophic level: grazers
shredders
collector-gathers
nectar feeders

2nd trophic level: predators

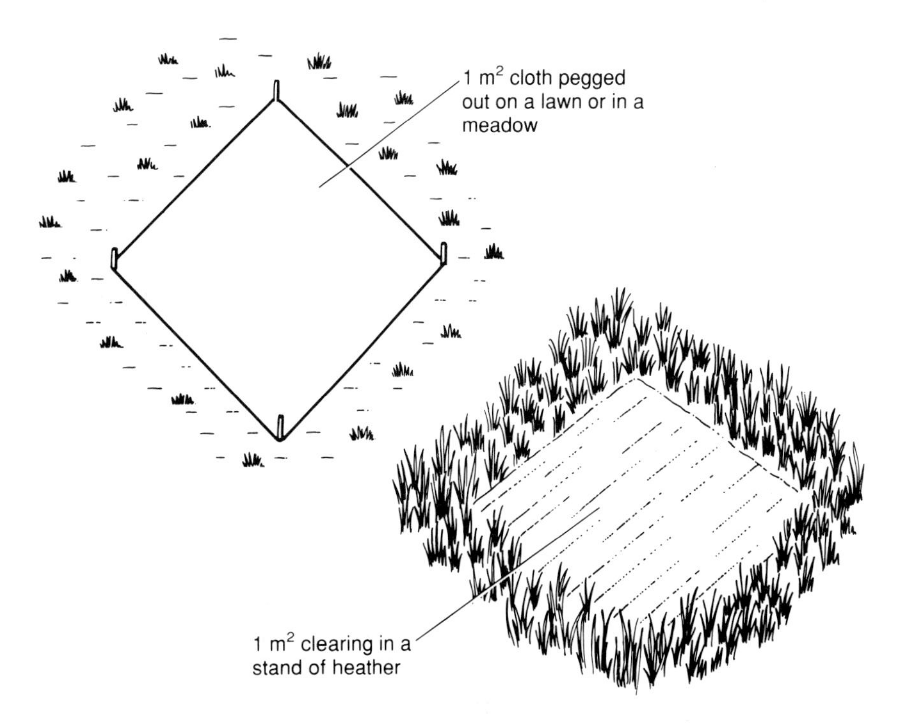

Figure 1 ***Estimating the size of insect populations by direct counts in the field***

investigation 19

Sample size and total population estimates

Estimates of total animal population (N) in the field can be made by using mark-release-recapture procedures. One of the most widely used of these procedures is known as the **Lincoln Index**, which can be expressed by the following general equation:

$$\text{total population (N)} = \frac{\text{no. individuals marked and released (1st sample)} \times \text{no. individuals in 2nd sample}}{\text{no. marked individuals in 2nd sample}}$$

In this investigation, carried out in the laboratory, the aim is to illustrate the effect of sample size on estimates of a known total population of 100 individuals. Results can be plotted as a graph to illustrate the effect of sample size on estimates of N.

Equipment and materials

- polythene bag ('No. 1'), containing 100 white paper squares or discs
- polythene bag ('No. 2'), containing 40 black paper squares or discs
- graph paper

ANIMAL ECOLOGY

Experimental procedure

1 Remove five white squares or discs from bag No. 1 and replace then with five black squares or discs from bag No. 2. (You now have a total population of 100 individuals of which 5% are marked.)

2 Shake the contents of bag No. 1. With your eyes closed, or without looking, now draw out 20 squares or discs from bag No. 1. Estimate the size of the total population in bag No. 1 by substituting your results into the following equation:

$$\text{total population (N)} = \frac{\text{no. black squares or discs in population (5)} \times \text{no. squares or discs recaptured (20)}}{\text{no. black squares or discs in 'recaptured' sample}}$$

Repeat this procedure to obtain a second estimate. Calculate a mean (average) value.

3 Repeat procedure **2** when bag No. 1 contains respectively 10, 15, 20, 25, 30, 35 and 40% black squares or discs. Record your results.

4 Present your results, including mean values, in the form of a table.

5 Plot your results as a graph, showing the predicted size of the total population against the percentage of marked individuals (see Figure 1). In your view, what percentage of this population needs to be marked before useful estimates of total population size can be obtained?

6 Calculate the standard error of your population estimates from the following equation:

$$\text{standard error} = N \times \sqrt{\frac{1}{(Rm)} - \frac{1}{(Rm - Ru)}}$$

where N = population estimate
Rm = number of marked individuals
Ru = number of unmarked individuals

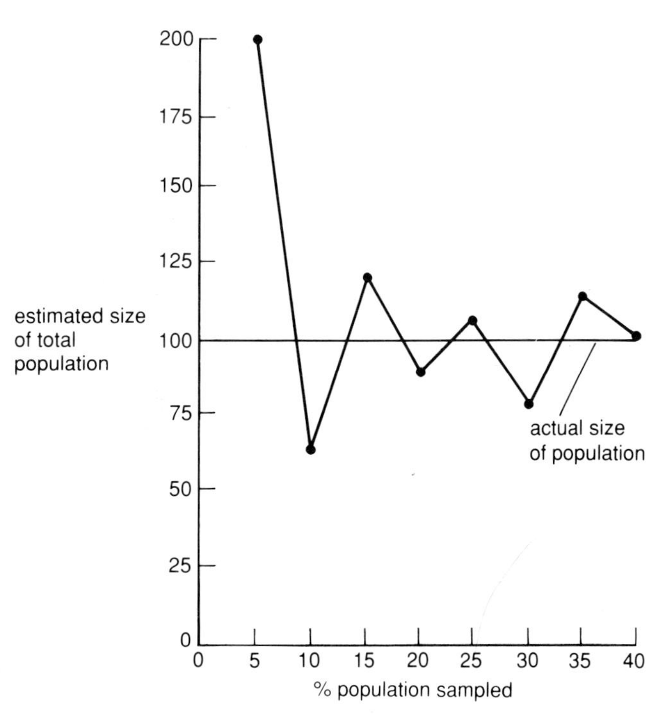

***Figure 1** The relationship between sample size and estimates of total population*

Taking it further

1 Apply a mark-release-recapture procedure to estimate the size of an animal population in the field. Select pond insects, land beetles, snails or other small invertebrates. Mark them with quick-drying paint in a position that does not (a) make them more conspicuous to their predators or (b) interfere with their locomotion.

2 Starting with a population of 500 paper discs, find the accuracy of predictions of total population size based on sampling 5, 10, 15, 20, 25, 30, 35 and 40% of the individuals. How do these results compare with those obtained with a total population of 100 individuals?

3 Cut 250 white paper discs with a hole punch. Prepare a piece of cardboard, 10 × 10 cm, and place 50 paper discs on it, distributed at random. Take a pin, close your eyes, and touch the cardboard. Repeat this 50 or 100 times and record the number of direct hits when the density of paper discs is 100, 150, 200 and 250. Plot a graph to show the number of direct hits against the density of the discs. In what ways does this exercise simulate the effects of a predator on a population?

investigation 20

Population growth in mealworm beetles

Studies of population growth can be carried out using mealworm beetles (*Tenebrio molitor*), supplied either with different amounts of the same food or different types of flour, meal or ground cereal. Kept in large Kilner jars, at temperatures of 23–30°C, these beetles will complete a life cycle within about 4 months (Figure 1a). Furthermore, they require very little attention. A small segment of fresh apple, or thin slice of carrot, placed into each jar at regular monthly intervals, provides the adults and larvae with all the moisture they require. Counts of the numbers of larvae, adults and pupae can be made at monthly intervals by passing the contents of each jar through a coarse sieve.

Equipment and materials

- 400 small mealworm beetle larvae (*T. molitor*, size = 1.0–1.5 cm)
- 8 × 4–8 lb Kilner jars
- 1 kg white flour
- 1 kg brown flour
- 1 kg bran
- 1 kg cornflour
- 1 kg semolina
- 1 kg ground rice
- 1 kg crushed cornflakes
- 1 kg crushed dog biscuit
- coarse sieve
- plastic jam pot covers or aluminium foil
- elastic bands
- glass-marking pen

ANIMAL ECOLOGY

Experimental procedure

The apparatus is shown in Figure 1b.

1 Label the Kilner jars from 1 to 8. Put each type of food into a separate Kilner jar, and write the name of the food on each jar.

2 Introduce 50 mealworm beetle larvae into each jar.

3 Cover each jar with a plastic jam pot cover or aluminium foil. Fit an elastic band to hold the cover or foil in position.
Place the jars in a warm place, preferably within the temperature range 25–30°C.

4 At monthly intervals, open the jars, sieve the contents and make counts of the numbers of larvae, pupae and adults present. Record all your results. Before sealing the jars again, introduce a small piece of apple or carrot.

5 After 6, 12 or 24 months, draw graphs to show changes in the numbers of larvae, pupae and adults in each jar. What do you conclude? List the different foods in rank order, according to their effects on population density. Start with the one in which population density reached the highest peak.

Taking it further

(a) *Stages in life cycle*

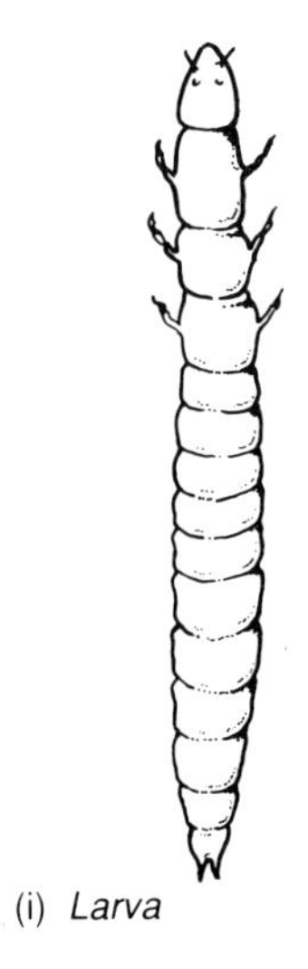

(i) *Larva*

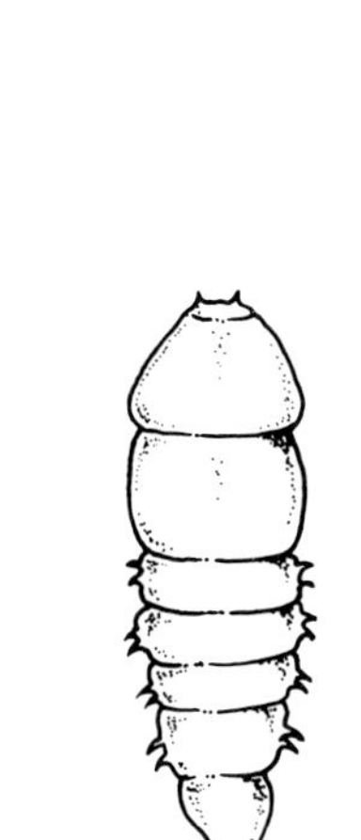

(ii) *Pupa*

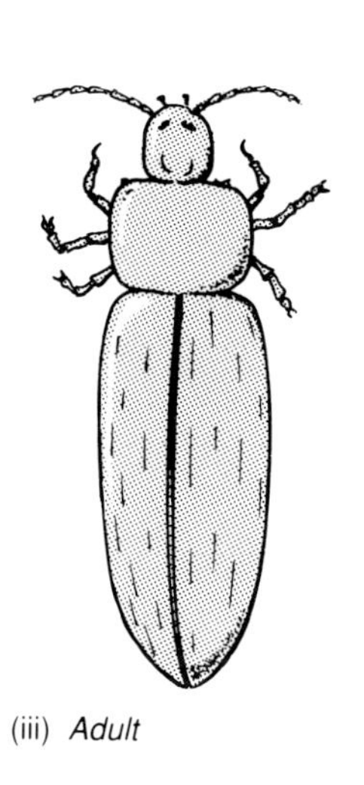

(iii) *Adult*

1 Devise and set up an investigation to find out how different masses of brown flour or bran affect subsequent changes in the numbers of mealworm beetles, from an initial population of 50 larvae per culture.

2 Mealworm beetle larvae and woodlice show different responses to changes in the relative humidity of the air. In both animals, the primary response consists of changes in the rate of locomotion.

Take 12 petri dishes and pour 25 cm^3 of each of the following saturated salt solutions into two dishes:

dish number	saturated salt solution	approximate % relative humidity at 20°C
1, 7	LiCl	12.5
2, 8	$MgCl_2\ 6H_2O$	33.0
3, 9	KNO_2	48.5
4, 10	NaCl	76.0
5, 11	KCl	85.0
6, 12	KNO_3	93.5

Label the dishes from **1–6** (mealworms) and **7–12** (woodlice).

Place small pieces of 'Blutac' around the inside wall of each dish, just above the level of the saturated solution. Use these as supports for a platform of plastic gauze, cut carefully to fit into the base of each dish. Finally, put a mealworm beetle larvae into each dish numbered 1–6 and a woodlouse into each dish numbered 7–12. Replace the lids and allow 20 minutes for equilibrium to be established. Place the petri dishes on the bench surface and set up an illuminated electric light bulb above them. Both animals will move around the circumference of the dish maintaining close contact with the edge. Record the time taken for each animal to complete one circumference of the dish (πd). Record the rates of locomotion above each saturated solution and relate them to the % relative humidity of the air.

(b) *Culture jar*

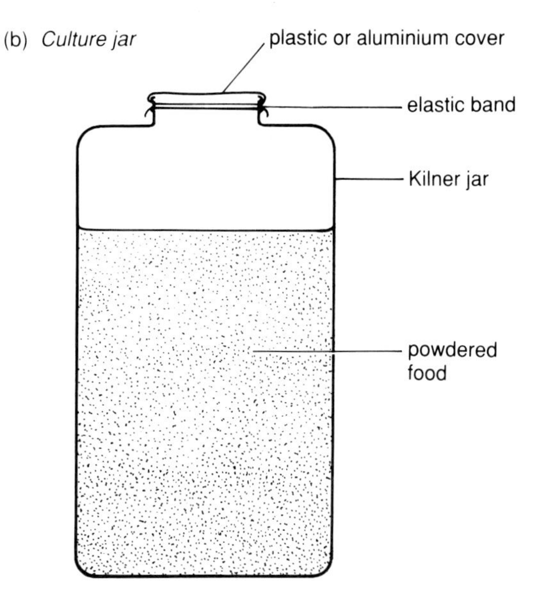

Figure 1** **Investigating population growth in **T. molitor**

investigation 21

Estimating animal populations in soil

Many different types of invertebrates, including arthropods, annelids and molluscs, live in the soil. Arthropods can be extracted and counted by using a Tüllgren funnel (Figure 1). This apparatus contains an electric light bulb, the source of both light and heat, positioned above a sieve of zinc or plastic gauze. Samples of soil are placed on the gauze and illuminated. As the soil gets warmer and drier, the arthropods move further and further down into the soil to get away from the light and heat source. When they finally reach the gauze they fall through the holes into a collecting vessel containing preservative.

Keys can be used to identify the species that have been collected. Estimates can be made quantitative by extracting arthropods from unit masses of soil.

Equipment and materials

- woodland soil
- garden trowel
- Tüllgren funnel
- 50% w/v methanol (or ethanol) in a collecting vessel
- plastic bags and string
- ruler
- top pan balance
- keys for identification of insects and other arthropods
- glass-marking pen

Experimental procedure

1 Select a woodland site with a deep surface layer of leaf litter.
2 Dig soil samples from (a) the soil surface (b) 5 cm below the surface and (c) 10 cm below the surface. Put each sample into a plastic bag, tie it with string, and label it.
3 On returning to the laboratory, weigh out 250 g surface soil and transfer it to the Tüllgren funnel. Switch on the lamp, and allow at least 24 hours for all the arthropods to pass through the soil into the container (Figure 1).
4 Repeat procedure **3** with soil collected from (a) 5 cm below the surface and (b) 10 cm below the surface.
5 Use keys to identify the arthropods contained in the samples. Prepare a table, similar to the one shown below, and enter the names of species wherever possible.

Type of arthropod	No. arthropods in samples		
	Surface soil	5cm below surface	10cm below surface
Insects a) adults b) larvae			
Crustaceans			
Arachnids a) spiders b) harvestmen c) mites			
Myriapods a) centipedes b) millipedes			

6 Summarise your results and draw conclusions.

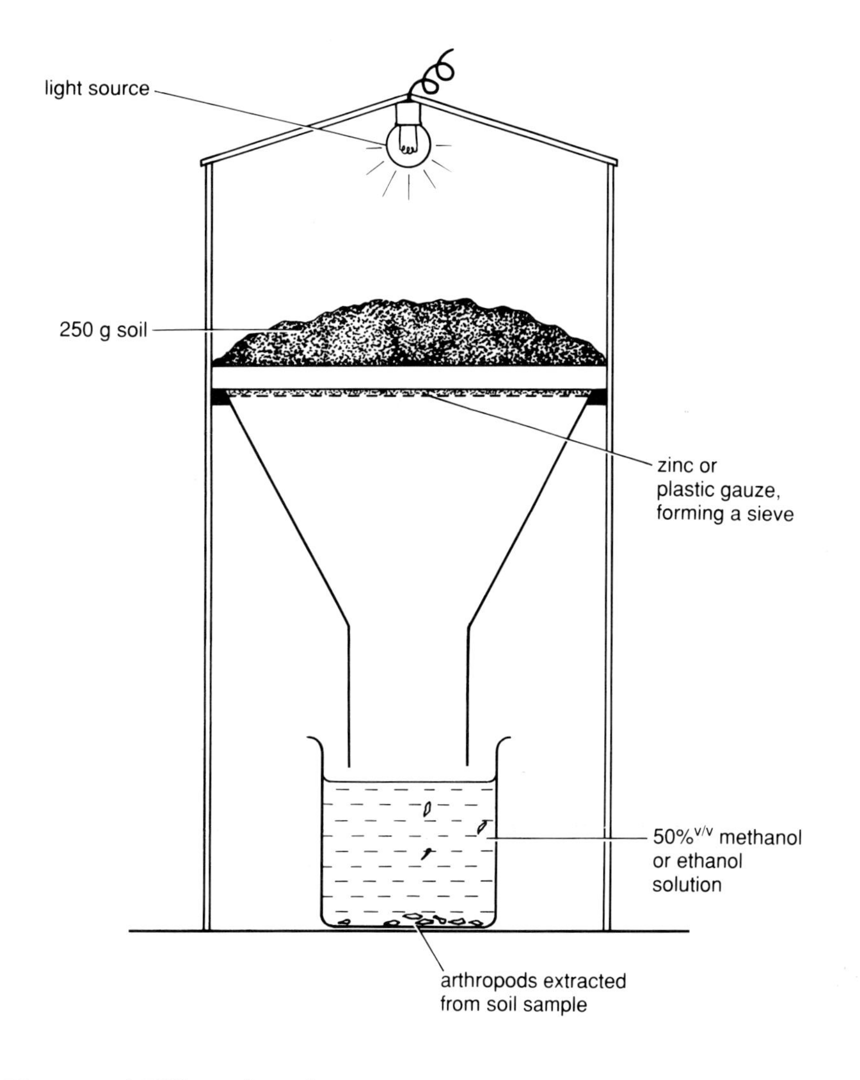

Figure 1 A Tüllgren funnel

Taking it further

1 Estimate the approximate number of earthworms in soil by placing 1 m^2 quadrats on the soil surface. Pour 2 dm^3 methanol over each 1 m^2 area and wait for the earthworms to rise to the surface. As they emerge, immediately wash them in a bucket of clean water, and retain them for counting. Compare numbers in (a) garden soil, (b) lawn and (c) meadow containing cows. Attempt an explanation of your results.

2 Estimate the approximate number of nematode worms in garden soil by tying 250 g soil in a muslin bag. Stand the bag in a 400 cm^3 beaker and add 50 cm^3 distilled water. After 24 hours use a syringe to withdraw 1 cm^3 of the soil solution. Place on drop of this solution on a microscope slide and count the number of nematode worms it contains. Find out the number of drops in 1 cm^3 soil solution then calculate the approximate number of nematodes in the soil sample.

Bibliography

Bayne, D, Evans, D, Llewellyn-Jones, J and Shalders, J (1988) *Biokeys* Blackie

Lewis, T and Taylor, L R (1967) *Introduction to Experimental Ecology* Academic Press.

investigation 22

Soil and pH preferences in earthworms

Earthworms often occur in large numbers in soils beneath oak and other broad-leaved trees. In these soils, dead organic material is rapidly broken down and mixed with mineral particles, a process in which earthworms play a major part. Conversely, earthworms are usually scarce in acid soils (below pH 5.0) typically found beneath conifers, heathers and rhododendrons. In this investigation the responses of earthworms to pH, their preference for soil types, and some of the factors affecting their rate of locomotion are determined.

Equipment and materials

- 20–30 small earthworms (e.g. 5–8 cm in length)
- buffered pH tablets (e.g. 4.0, 5.0, 6.0, 7.0, 8.0 and 9.0)
- 1 dm^3 beaker, containing distilled water
- clay
- sand
- loam
- humus (e.g. well rotted farmyard manure, compost)
- perspex lunch box with lid
- sheets of A4 paper
- sheets of sand paper (20 × 5 cm)
- sheet of plastic (20 × 5 cm)
- 2 × pieces of wood (e.g. 2 × 2 × 20 cm)
- 6 × 100 cm^3 beakers
- bench lamp, fitted with 40 W bulb
- forceps
- paper towels
- stop watch, or clock with a second hand
- spoon
- glass-marking pen

Experimental procedure

Soil preference

1 Take the lunch box and fill it to within 2–3 cm of the top with bands of sand, clay, loam and compost, as shown in Figure 1a. Add water to make the soil moist, but not sticky. Place 20–30 earthworms into the box and replace the lid. Stand the box in a dark place for at least 24 hours. Use a spoon to dig out each band of soil. Count the number of earthworms in each band and record your results.

Response to light and rate of locomotion

2 Place a sheet of A4 paper on the bench. Arrange the two pieces of wood so that they are parallel to one another and 0.5–1.0 cm apart. Switch on the bench lamp and place it about 15 cm from one end of the piece of wood. Place an earthworm at one end of the corridor between the pieces of wood, facing away from the light. Record the time taken for the earthworm to move through the corridor and emerge at the other end. Take a second reading and estimate a mean value. Calculate the rate of locomotion in cm/hr.

3 Repeat procedure **2** with (a) sandpaper and (b) plastic beneath the pieces of wood. Record all your results. Try to give an explanation for these results.

Response to pH

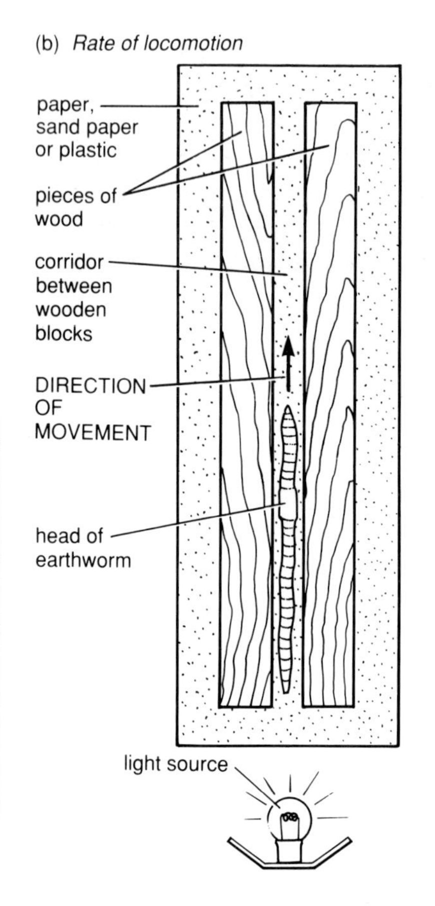

4 Prepare buffered pH solutions by dissolving tablets in distilled water according to the manufacturer's instructions. Pour about 80 cm^3 of each solution into a 100 cm^3 beaker. Label the beakers.

5 Place a piece of A4 paper beneath the pieces of wood, arranged as shown in Figure 1b. Switch on the light. Record the time taken for an earthworm to travel along the corridor. Dip this earthworm into buffered pH 4.0 solution for 10 seconds, then record the time it takes to move along the corridor. Take a second reading and estimate a mean value. Wash the earthworm in distilled water after you have measured its rate of locomotion.

6 Repeat procedure **5** at pH 5.0, 6.0, 7.0, 8.0 and 9.0. Record the time taken for the earthworm to move along the corridor at each pH. Apply a statistical test to find out if any differences you record are significant.

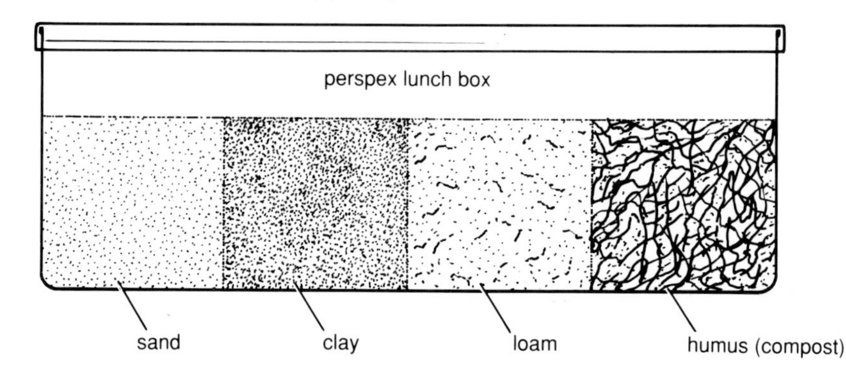

***Figure 1** Investigating the behaviour of earthworms*

Taking it further

1 Obtain leaf extracts from four of the following plants: holly, rhododendron, cupressus, pine, heather, oak, beech, sage, lavender, mint, lily of the valley. Paint circles of roughly 4 cm diameter on A4 paper with each extract. Place a small earthworm at the centre of each circle, before the extract has dried. Which extracts, if any, keep the earthworm confined within the circle? What is the significance of this behaviour?

2 Try to find out if earthworms prefer oak, beech or alder leaves as food sources. Both fresh and dried leaves, ground in a kitchen mixer, can be used in the investigation. Place weighed samples of food on squares of plastic, above moist loam in a lunch box. Use 20–30 earthworms in the investigation, and weigh the food samples every day. Keep the box in a dark place to encourage feeding.

investigation 23

Food preference in mealworm beetles

All animals form a part of one or more food chains, a linear sequence of eating and being eaten. Practical work, designed to find out the constituents of an animal's diet, may use one of two approaches. The first consists of dissecting out the gut contents of a freshly-killed animal and identifying the species that are undergoing digestion. An alternative approach consists of observing the feeding habits of living animals kept inside laboratory cages.

In this investigation mealworm beetles (*Tenebrio molitor*) are kept in translucent perspex lunch boxes and offered a variety of dried, powdered food products. Preferences are worked out by measuring the mass of each type of food before and after feeding.

Equipment and materials

- 40 × adult mealworm beetles (or larvae)
- 2 × perspex lunch boxes
- 6 × small watch glasses
- 3 g white flour
- 3 g brown flour
- 3 g bran
- 3 g semolina
- 3 g ground rice
- 3 g dried potato
- top pan balance
- glass-marking pen

Experimental procedure

1 Number the lunch boxes 1 and 2.
2 Place each type of food on a separate watch glass. Transfer these to the perspex lunch boxes, three per box.
3 Carefully place 20 mealworm beetles into each box, taking care not to disturb the food samples. Replace the lids of boxes. Stand the boxes in a warm place (e.g. at 16–23°C).
The apparatus is shown in Figure 1.
4 After 1–2 days remove each watch glass and weigh the food that it contains. Calculate the amount of each food that has been eaten.
5 Plot a histogram to show the mass of each food that the mealworm beetles have eaten.

Taking it further

1 Devise and carry out an investigation to find out if mealworm beetle larvae have (a) the same food preferences as the adults and (b) consume food at the same rate.

2 How would you investigate the claim that mealworm beetle larvae 'prefer damp flour, containing yeast'. What controls, if any, would be required?

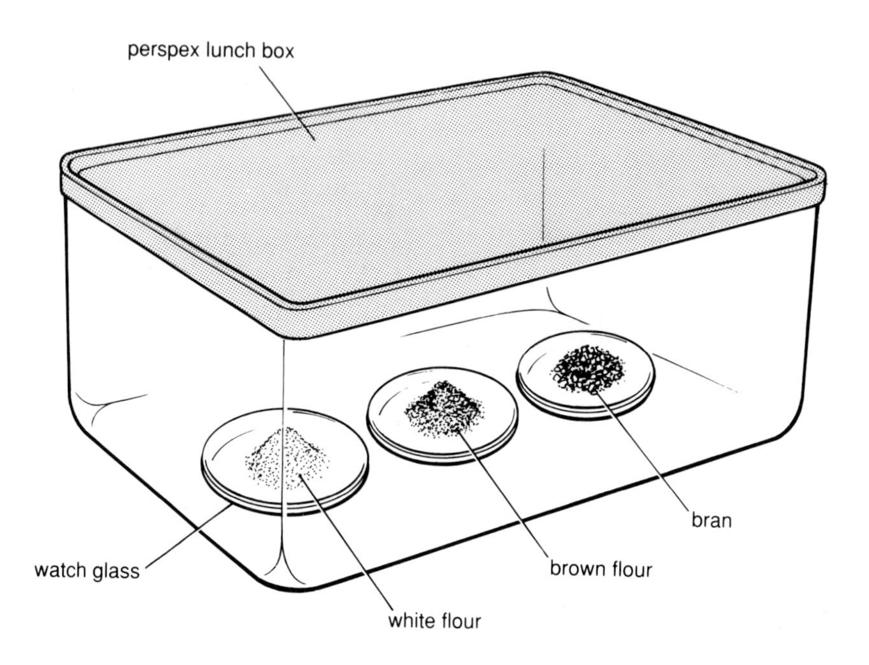

Figure 1 Apparatus for determining the food preferences of mealworm beetles

investigation 24

Behavioural responses of invertebrates

Invertebrates including insects, woodlice, spiders and mites, show a number of tactic or kinetic responses when subjected to directional external stimuli. These responses determine, in part, the habitat range of invertebrates and in many cases their preference for particular micro-habitats. Figure 1 shows how a boiling tube can be used to create simple, inexpensive choice chambers in which the behaviour of small animals can be observed. By using this apparatus, four different responses can be investigated.

Type of response	Nature of stimulant
Phototaxis or photokinesis	Light source
Hydrotaxis or hydrokinesis	Moisture gradient in air
Chemotaxis	Chemical substance in solution
Geotaxis	Gravity

In each case animals may move directly towards the source of stimulation (positive taxis) or directly away from it (negative taxis). Alternatively, their rate of locomotion may be changed by the intensity of the stimulus (kinesis), or they may fail to make any response (neutral response).

Equipment and materials

- 10 invertebrate animals (e.g. woodlice, beetles, blowfly maggots)
- dilute hydrochloric acid (e.g. 0.1 N)
- dilute sodium hydroxide solution (e.g. 0.1 N)
- milk
- 1% solution of 'Bovril' or 'Marmite'
- seven boiling tubes, fitted with rubber bungs
- sleeve of black paper (to fit one half of a tube)
- silica gel (six pieces in a piece of nylon stocking)
- elastic band
- cotton wool
- filter paper
- plasticine
- forceps
- scissors
- bench lamp, fitted with a 40 W bulb
- glass-marking pen

Experimental procedure

Response to light

1 Cut a sleeve of black paper or plastic to wrap around one half of a boiling tube. Use an elastic band to hold the sleeve in position. Put 10 invertebrate animals into the tube and fit a bung. Lay the apparatus on the bench surface, in a horizontal position, at a distance of 10 cm from the illuminated 40 W electric light bulb, and at right angles to it. At intervals of 30 seconds, over a period of 5 minutes, record the number of invertebrates in the illuminated and darkened halves of the apparatus. What is the response of the animals to light? How might this help the animal to survive in its natural habitat?

Response to a moisture gradient

2 Tie six or more granules of silica gel in a small square of nylon stocking. Place this into the bottom of a boiling tube. Lay the tube in a horizontal position on the bench surface and place 10 invertebrates into the tube. Place moist cotton wool into the neck of the tube before fitting the bung. Use the pen to mark a boundary between the dry and moist halves of the apparatus. At intervals of 1 minute, over a period of 5 minutes, record the number of animals in the dry and moist halves of the apparatus.

What response was made to water? How might this response help the animal to survive in its natural surroundings?

Response to chemical compounds

3 Cut four strips of filter paper, each approximately 4 × 7 cm. Dip one of these pieces into each of the following:

a) dilute hydrochloric acid
b) dilute sodium hydroxide solution
c) milk
d) 1%$^{w/v}$ solution of 'Bovril' or 'Marmite'.

Take four boiling tubes. Use forceps to place one piece of filter paper into the upper half of each tube leaving only a small gap through which the behaviour of the animals can be observed. Place 10 invertebrates into each tube. Replace the bungs and mark the mid-points of the tubes. Place the tubes on the bench surface, in a horizontal position. At intervals of 1 minute, over a period of 5 minutes, record the number of animals in each half of the tube. Are the responses to these compounds positive or negative? What is the significance of each response? How could the design of the apparatus be improved?

Response to gravity

4 Cut a strip of filter paper 3 cm wide and approximately 1 cm less than the full length of the boiling tube. Place the paper into the boiling tube and moisten it, so that it sticks to the glass. Place 10 invertebrates into the tube, and fit a bung. Make a mound of plasticine on the bench surface and support the tube as shown in Figure 1d, inclined at an angle of approximately 10° to the bench surface. Mark the mid-point of the tube. At intervals of 1 minute, over a period of 5 minutes, record the number of invertebrates in the upper and lower halves of the tube.

What is the response of the animals to gravity? How may this response assist the animal's survival in its natural habitat?

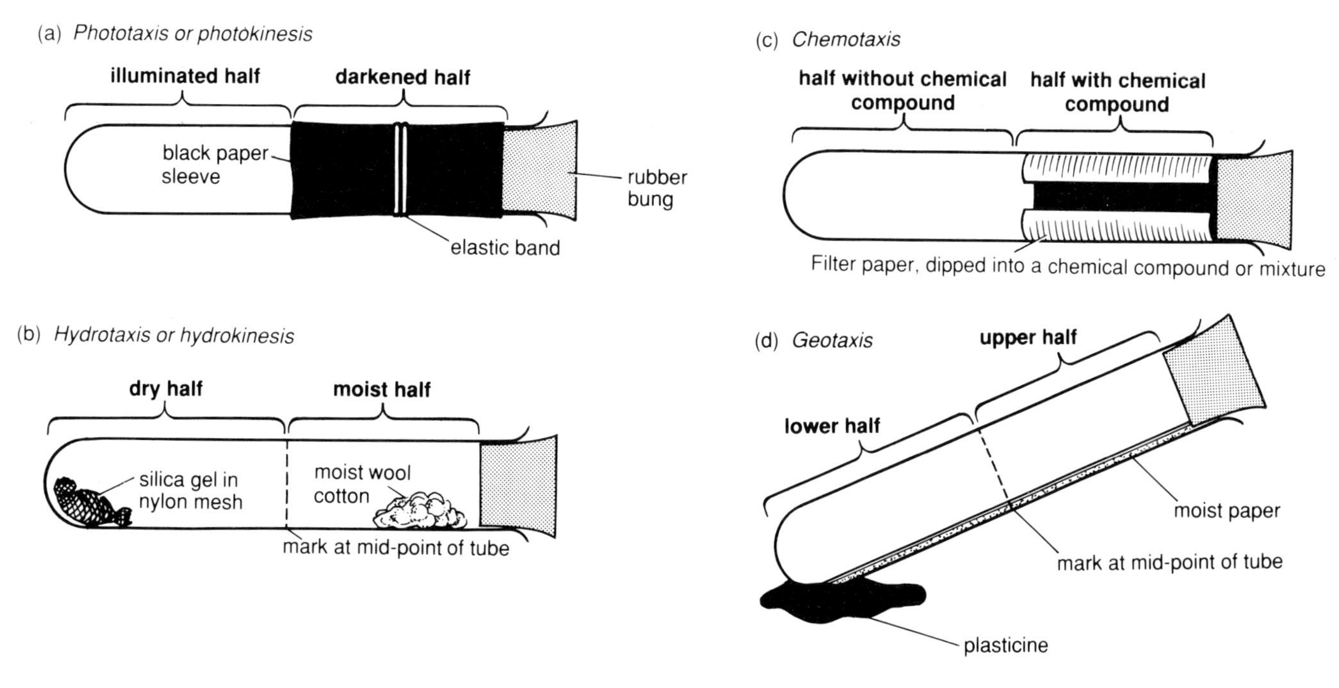

Figure 1 Apparatus for investigating the behavioural responses of invertebrates

Taking it further

Simple choice chambers can also be made from plastic petri dishes with three partitions, available from suppliers (Figure 2). Working in a fume cupboard, use a red hot bradawl or wire to burn holes in each of the three partitions, close to the mid-point, so that invertebrates can move between compartments. Place moist cotton wool in one compartment, and silica gel (tied in nylon stocking) in another. Use 9 or 12 animals in each dish. Analyse your results statistically using the χ^2 test for two variables with expectation. The null hypothesis is that the animals are as likely to choose one compartment as any other, i.e. you would expect to find equal numbers of animals in each compartment (see Dowdeswell p118).

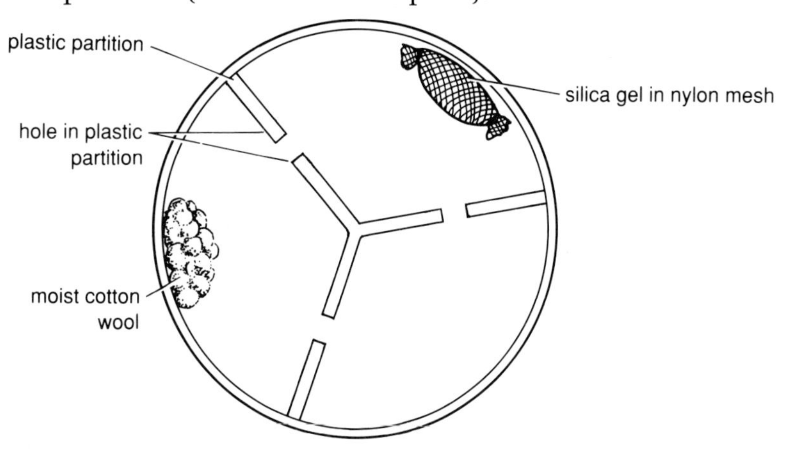

Figure 2 Three-compartment choice chamber made from a petri dish

Bibliography

Dowdeswell, W H (1959) *Practical Animal Ecology* Methuen, London.

Aggregation in arthropods

Small arthropods, including many insects, have a relatively large surface area : volume ratio. They are therefore in constant danger of losing water and may become desiccated following prolonged exposure to high temperatures, low relative humidity, or high wind velocity. As a means of reducing water loss from exposed surfaces, many arthropods place a part of their body in contact with a solid object, a response known as **positive thrigmotaxis.** Such animals frequently aggregate, as contact with one another serves a similar purpose. In this investigation, simple apparatus is used to find out if selected species of arthropod (a) show a positive thrigmotactic response and (b) tend to aggregate.

Equipment and materials

- 16 arthropods (e.g. beetles, woodlice, blowfly maggots)
- 2 × white tiles
- 2 × 9 cm plastic petri dishes
- stop watch, or clock with a second hand
- compass
- ruler
- glass-marking pen

Experimental procedure

1 Using the compass and marking pen, draw circles on the base of a petri dish at 2 cm and 4 cm from the margin. Place a white tile on the bench surface. Invert the base of the petri dish, and place it on the tile, as shown in Figure 1a. Place 10 animals beneath the inverted dish. At intervals of 30 seconds over a period of 5 minutes, count the number of animals in the inner, middle and outer circles. Record and tabulate your results. What do you conclude?

2 Using a marking pen and ruler, draw lines across the tile from corner to corner, and between the mid-points of opposite sides. This will divide the tile into eight segments, which can be numbered. Invert the base of the second petri dish over the centre of the tile, as shown in Figure 1b. Place 16 arthropods beneath the inverted dish. At intervals of 30 seconds, over a period of 5 minutes, count the number of animals in each segment of the dish. Record your results in the form of a table. Plot a graph of your results. What do you conclude?

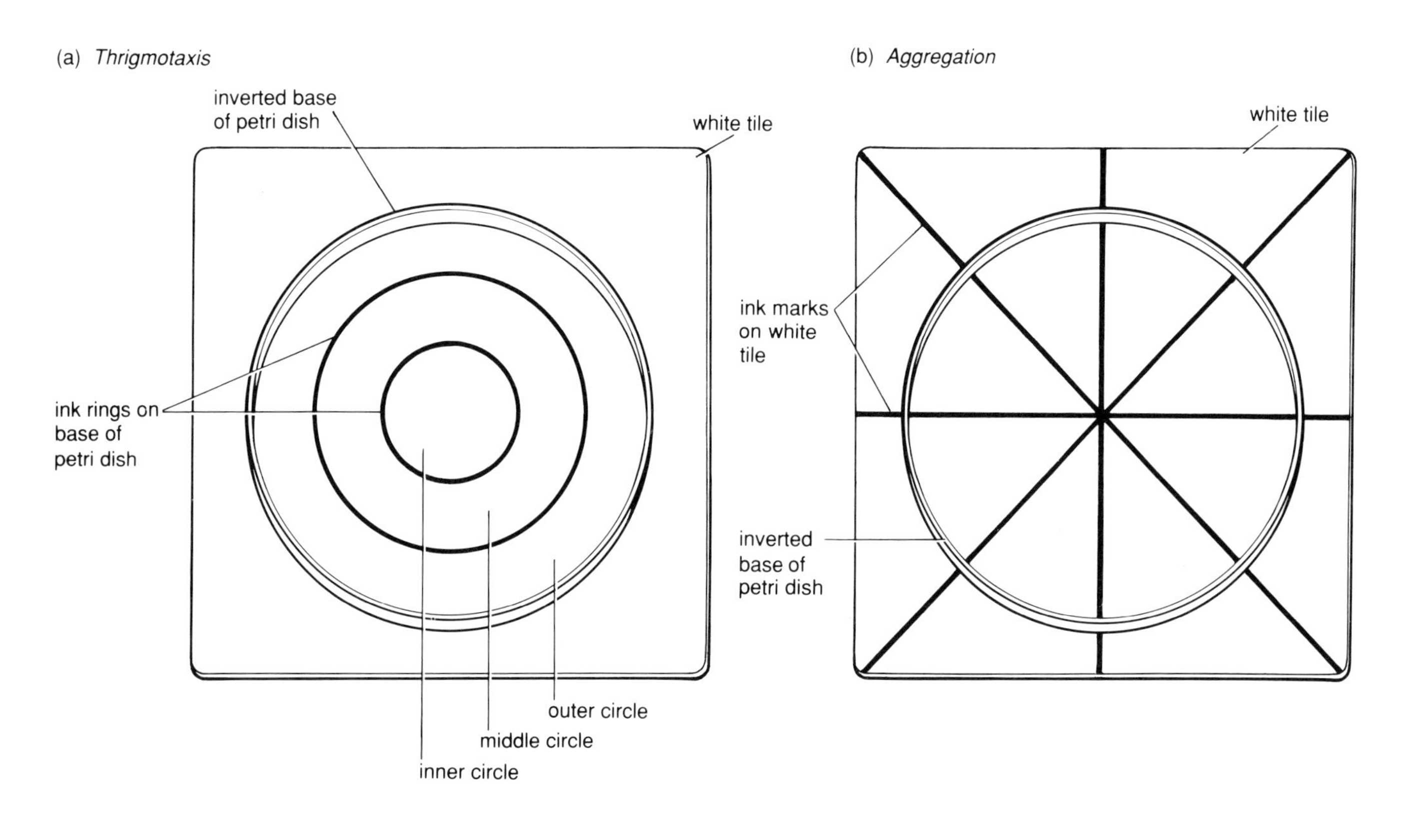

Figure 1 Apparatus for investigating thrigmotaxis and aggregation in arthropods

Taking it further

Rheotaxis is a response to water currents, shown by some animals that live in streams and rivers. Animals that show a positive rheotactic response position their bodies so that they are facing the source of the current. Figure 2 shows a simple piece of apparatus, constructed from a plastic lunch box and tubing, that can be used to investigate the rheotactic response and to measure the flow rates of water currents which can be tolerated by different species. Construct the apparatus and use it to investigate the behaviour of 5 or more different stream-dwelling animals.

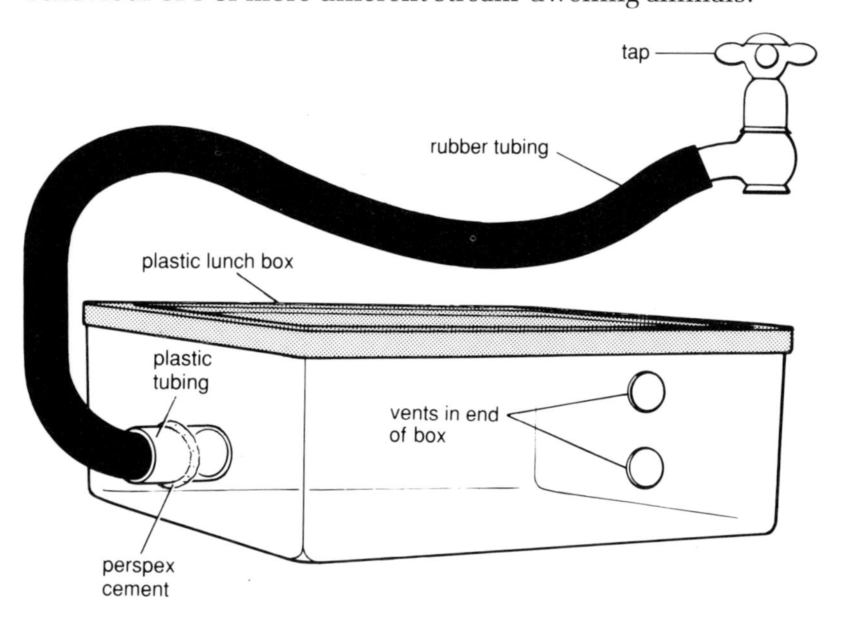

Figure 2 Apparatus for determining the effect of water currents on stream-dwelling invertebrates

Organic pollution of water samples

Water in ponds, lakes and rivers is often polluted by organic matter, such as leaf litter and dead animals. The presence of this material in the water causes an increase in the number of micro-organisms. Populations of bacteria and fungi are the first to increase, followed by populations of microscopic animals and algae. The more organic matter in the water, the more micro-organisms it usually contains.

All of these micro-organisms produce the enzyme **catalase** which rapidly breaks down hydrogen peroxide into oxygen and water:

$$\underset{\text{(liquid)}}{\text{Hydrogen peroxide}} \xrightarrow{\text{catalase}} \underset{\text{(gas)}}{\text{oxygen}} + \underset{\text{(liquid)}}{\text{water}}$$

The rate at which oxygen is released depends on the number of micro-organisms in the water. Water that is heavily polluted by organic matter will therefore break down hydrogen peroxide more rapidly than unpolluted water. In this investigation you will mix equal volumes of hydrogen peroxide solution and water, and measure the rate at which oxygen is released. If bubbles of oxygen are trapped in the barrel of a syringe, an equivalent volume of the mixture is displaced into a capilliary tube, where it can be measured.

Equipment and materials

- 10 cm^3 pond water
- 10 cm^3 tap water
- 5 cm^3 5-vol hydrogen peroxide solution
- 2 × 1 cm^3 plastic syringes
- 2 × 2 cm lengths of capillary tubing
- 2 × 2 cm lengths of rubber tubing
- boss, clamp and retort stand
- stop-clock, or watch with a second hand
- ruler, graduated in mm
- glass-marking pen
- eye protection
- plastic gloves

Experimental procedure

1 Put on your eye protection and plastic gloves. Number the syringes 1 and 2. Draw hydrogen peroxide solution into syringe 1 to the 0.5 cm^3 mark. Draw tap water into the same syringe until the mixture reaches the 1.0 cm^3 mark. Gently rock the contents of the syringe to assist mixing. Attach rubber tubing to the nozzle of the syringe, and fit a capillary tube, as shown in Figure 1. Support the apparatus in a clamp, at about 25 cm above the bench surface.
2 Set up syringe 2 with 0.5 cm^3 hydrogen peroxide solution and 0.5 cm^3 pond water.

3 Apply gentle pressure to the handle of each syringe until a meniscus appears at the top of each capillary tube. Mark the position of each meniscus.
4 At intervals of 1 minute, over a period of 5–20 minutes, mark the position of the meniscus in each capillary tube. Record your results.
5 Draw a graph of your results.

Taking it further

1 Design and carry out an investigation to find out if water from the (a) top or (b) bottom of a pond is most heavily polluted by organic matter. Write out your method, listing instructions in the order they should be carried out.
2 Artificial sea water can be prepared by dissolving tablets, containing mineral salts, in distilled water. Design and carry out an investigation to find out if samples of sea water, collected from holiday beaches, are polluted.
3 Devise and carry out an investigation of relative population growth rates in two microscopic pond-dwelling algae. Use catalase production as a means of comparing the relative numbers of individuals in unit samples of pond water.

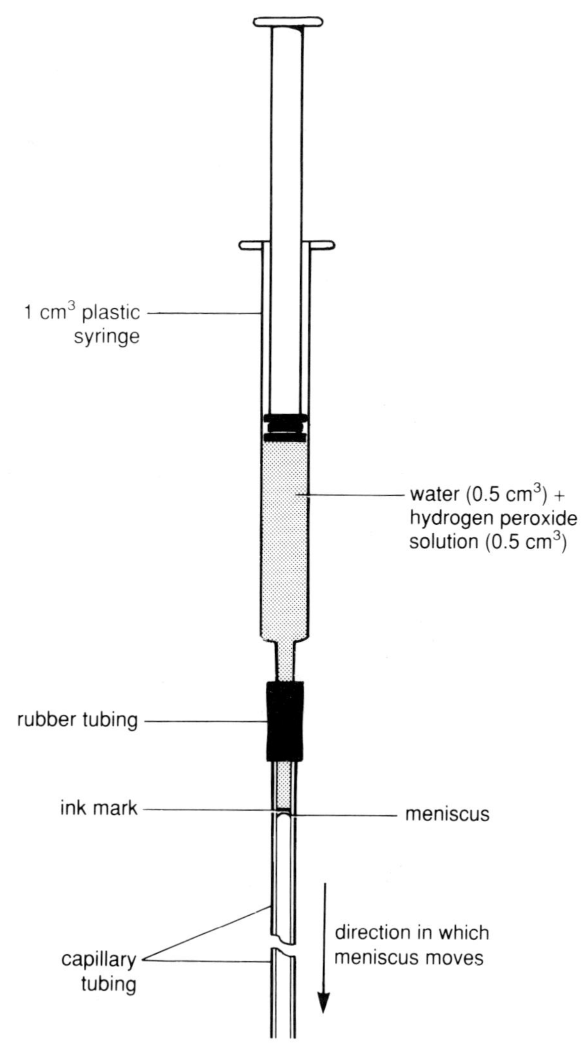

Figure 1 Apparatus for determining relative pollution levels in water samples

Air pollution and fungal populations

Many saprophytic fungi grow on leaves, especially on their lower or abaxial surfaces. Most are ascomycetes that form yellow, pink, brown or white smooth gelatinous colonies when grown on malt or nutrient agar. Others are zygomycetes, composed of fine branching threads or hyphae. These form grey-white colonies, often resembling tufts of cotton wool. The population density of all these fungi, normally highest in late summer, is affected by air pollution. Acid pollutants such as sulphur dioxide and nitrogen dioxide reduce fungal populations on the surfaces of leaves. A similar effect is produce by minute amounts of lead from car exhausts. The degree of fungal inhibition can be used as an index of pollution. In general, the more polluted the air, the fewer fungi are present on the surfaces of leaves. In this investigation you will use two methods for estimating the density of fungi on the lower surfaces of leaves.

Equipment and materials

- leaves of a known tree, shrub or annual from a polluted area (e.g. roadside verge)
- leaves of the same tree, shrub or annual from a relatively unpolluted area (e.g. meadow)
- 4 × petri dishes containing malt or nutrient agar
- number 12 cork borer
- forceps
- 'Vaseline'
- 'Sellotape'
- incubator, maintained at 25°C
- glass-marking pen

Experimental procedure

1 Number the petri dishes from 1–4.
2 Remove the lid from dish 1. Use 'Sellotape' to fix a leaf from the polluted area, lower side down, inside the lid of the dish, as shown in Figure 1a. Return the lid and seal the dish. Tap the lid several times and incubate the dish at 25°C. After 1–2 weeks, coloured fungal colonies should appear on the surface of the agar.
3 Repeat procedure **2** with the leaf from the unpolluted area.
4 Use the cork borer to cut six discs from a leaf taken from each site. Smear Vaseline on the upper surface of each leaf disc. Use forceps to stick six discs from the polluted area to the lid of dish 3, and six discs from the unpolluted area to the lid of dish 4 (Figure 1b). Return the lids and seal the dishes. Tap the lids several times, and incubate the dishes at 25°C. Wait for the appearance of coloured fungal colonies.
5 Record the number and type of fungal colonies formed on each agar plate. Calculate the density per 1 cm^2 of leaf surface.

Taking it further

1 Extend your investigation to include direct leaf prints of both the upper and lower surfaces of leaves onto malt or nutrient agar. Firmly press the leaf against the agar to transfer any fungal spores from the surface of the leaf to the agar. Seal the dishes and incubate them at 25°C for 1–2 weeks. Without opening the dishes, apply a square of graph paper, in which a 1 cm^2 hole has been cut, to those areas of the dishes in which leaf prints were made (Figure 1c). Count the number of colonies visible through the hole. Make 3–5 counts per dish, then calculate mean values. Record all your results. Autoclave the petri dishes after use.

2 Examine the condition of the epidermis in leaves from polluted and unpolluted areas. Paint the surfaces with clear nail varnish. After the varnish has dried, peel it off and examine the imprint under a microscope. How do the imprints of leaves from polluted and unpolluted areas differ?

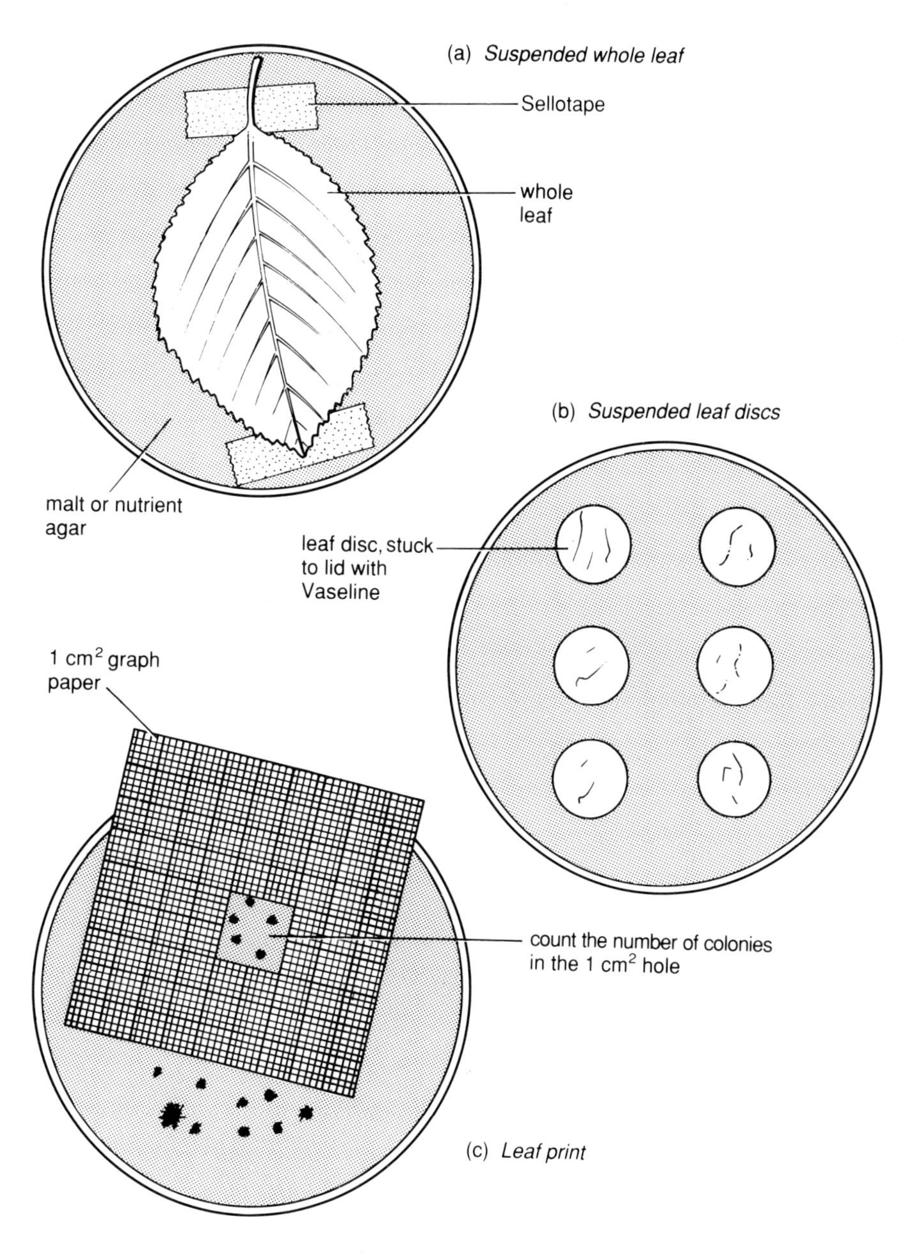

Figure 1 Methods of estimating density of fungal populations on leaf surfaces

Suppliers

1 Difco Laboratories
East Molesley
Surrey KT8 0SE
Resources: agars

2 Griffin and George Ltd
Gerrard House
Worthing Road
East Preston
West Sussex
BN16 1AS
Resources: laboratory chemicals, quadrats, light meters, temperature probes, oxygen meters

3 Irwin-Desman Ltd
294 Purley Way
Croydon
CR9 4QL
Resources: laboratory chemicals, multimeters

4 Philip Harris Biological Ltd
Oldmixon
Weston-super-Mare
Avon
BS24 9BJ
Resources: laboratory chemicals, quadrats, light meters, temperature probes, oxygen meters

5 Hughes and Hughes (enzymes) Ltd
Unit 1F
Lowmoor Industrial Estate
Tonedale
Wellington
Somerset
TA21 0AZ
Resources: enzymes, glass-marking pens

6 Oxoid Ltd
Basingstoke
Hampshire
RT24 0PW
Resources: agars

7 Solex International
Cottage Lane Industrial Estate
Broughton Astley
Leicestershire
LE9 6PD
Resources: light meters, digital thermometers, pH meters, water testers, weighing equipment

British Library Cataloguing in Publication Data

Freeland, Peter
Habitats and the environment: investigations.
– (Focus on biology)
I. Title II. Series
574.5

ISBN 0 340 55433 9

First published 1992
Impression number 10 9 8 7 6 5 4 3 2
Year 1999 1998 1997 1996 1995 1994

Typeset by Litho Link Ltd, Welshpool, Powys, Wales.
Printed in Great Britain for Hodder & Stoughton Educational, a division of Hodder Headline Plc, 338 Euston Road, London NW1 3BH by St Edmundsbury Press Limited, Bury St Edmunds, Suffolk.